T0341322

Fundamentals of Sustainability in Civil Engineering

Fundamentals of Sustainability in Civil Engineering

Second Edition

Andrew Braham
Sadie Casillas

CRC Press
Taylor & Francis Group
Boca Raton London New York

CRC Press is an imprint of the
Taylor & Francis Group, an **informa** business

Second edition published 2021
by CRC Press
6000 Broken Sound Parkway NW, Suite 300, Boca Raton, FL 33487-2742

and by CRC Press
2 Park Square, Milton Park, Abingdon, Oxon, OX14 4RN

© 2021 Taylor & Francis Group, LLC

First edition published by CRC Press 2017

CRC Press is an imprint of Taylor & Francis Group, LLC

Library of Congress Cataloging-in-Publication Data

Names: Braham, Andrew, author. | Casillas, Sadie, author.
Title: Fundamentals of sustainability in civil engineering / Andrew Braham, Sadie Casillas.
Description: Second edition. | Boca Raton, FL : CRC Press/Taylor & Francis Group, LLC, 2021. | Includes index.
Identifiers: LCCN 2020035617 (print) | LCCN 2020035618 (ebook) | ISBN 9780367420253 (hardback) | ISBN 9780367817442 (ebook)
Subjects: LCSH: Sustainable engineering--Textbooks. | Civil engineering--Environmental aspects--Textbooks.
Classification: LCC TA145 .B74 2021 (print) | LCC TA145 (ebook) | DDC 624.028/6--dc23
LC record available at https://lccn.loc.gov/2020035617
LC ebook record available at https://lccn.loc.gov/2020035618

ISBN: 978-0-367-42025-3 (hbk)
ISBN: 978-0-367-81744-2 (ebk)

Typeset in Times
by Deanta Global Publishing Services, Chennai, India

To our loving spouses, Brenda and Bryan

Contents

Preface

The content of this textbook targets a senior-level undergraduate course in Civil Engineering. This textbook is intended to introduce students to the broad concept of sustainability while also preparing them for the Fundamentals of Engineering (FE) exam. Effort has been made to utilize concepts from the FE reference manual so that students become familiar with or reacquainted with terminology and nomenclature utilized in the FE reference manual.

All attempts were made to remove any errors from this book. However, there are no doubt lingering issues here and there, and for that we apologize. If you find an error within the text, whether grammatical or technical, please do not hesitate to email Andrew at afbraham@uark.edu.

Thank you!

Andrew Braham and Sadie Casillas

Acknowledgments

We could not have completed this book on our own. Thank you to our colleagues who reviewed chapters in their subject areas, including Ken Sandler, Zola Moon, Wen Zhang, Richard Welcher, Gary Prinz, Kirk Morrow, and Eric Fernstrom. We also greatly appreciate our colleagues who helped us through polishing concepts and homework problems, including Julian Fairey, Findlay Edwards, and Rodolfo Valdes-Vasquez. At the end of the day, we would not be in our current position without students, and we thank all of the students who have taken the course *Sustainability in Civil Engineering at the University of Arkansas.* So many of the ideas and concepts have been taken directly out of our learnings in the classroom.

Thank you to all the staff at Taylor & Francis who were always quick to help and who made the process as easy as possible for me.

Finally, a huge thanks to our personal editors, Judy, Brenda, and Bryan. We could not have done this without you!

Authors

Andrew Braham is an associate professor of civil engineering at the University of Arkansas. He graduated from the University of Wisconsin with a BS in May 2000 and an MS in May 2002. From June 2002 through December 2004, he worked with Koch Materials Company as a field engineer and a research engineer. In June 2005, he returned to school for doctoral work at the University of Illinois and graduated in December 2008. For the next two years, from February 2009 to November 2010, Dr. Braham was a postdoctoral research fellow at Southeast University in Nanjing, China. In November 2010, Dr. Braham began work in the Civil Engineering Department at the University of Arkansas, where he has continued to study transportation materials while expanding into sustainability. A major component of Dr. Braham's laboratory investigates asphalt emulsion, pavement maintenance, and pavement rehabilitation. He has performed research for regional, state, and national funding agencies, along with private industry. He teaches undergraduate and graduate courses in transportation engineering, pavement design, pavement materials, and sustainability. You can find out more about Dr. Braham at www.andrewbraham.com.

Sadie Casillas is a research civil engineer at the US Army Engineer Research and Development Center. She received a BS from the University of Arkansas in May 2013 followed by an MS in May 2015. For the next year, she worked at HBK Engineering in Chicago, Illinois, as a civil design engineer for the telecommunications sector. In June of 2016, she returned to the University of Arkansas to pursue a PhD, which she completed in June 2020. Upon graduating, she began working for the US Army Engineer Research and Development Center in July 2020 in the Airfields and Pavement Branch. Dr. Casillas's research will focus on asphalt materials with an emphasis on pavement preservation and pavement rehabilitation. You can find more about Dr. Casillas on her LinkedIn page: www.linkedin.com/in/sadie-casillas.

1 Introduction to Sustainability

We hold the future in our hands, together, we must ensure that our grandchildren will not have to ask why we failed to do the right things, and let them suffer the consequences.

Ban Ki-moon

The term "sustainability" is currently very popular. Industries and organizations realize the benefits of protecting the future while succeeding in the present. In the present, sustainability is most often defined as incorporating three pillars into design: economics, environmental, and social. However, the general concepts of sustainability have been in use for millennia. The design and construction of Roman aqueducts for drinking water distribution were so robust that they have lasted centuries, with dozens of aqueducts built as early as 300 BC still standing today and the "Roman Road" still being used for movement of traffic. The Iroquois Native American confederacy has been in place since approximately the 12th century and uses the concept of sustainability in its constitution. Finally, today, there is a significant push for many sustainability initiatives, including more fuel-efficient vehicles on the roadways. Fuel-efficient vehicles address all three pillars of sustainability, by reducing fuel consumption (economics), decreasing emissions (environmental), and allowing more diverse transportation options for consumers (social or society). There are literally hundreds of existing books on sustainability discussing these and other concepts, but in order to demonstrate the development of sustainability overall, the United Nations will be used as an example to show how sustainability has been qualified and quantified over the past forty years.

1.1 THE DEVELOPMENT OF SUSTAINABILITY THROUGH THE UNITED NATIONS

The United Nations, or the UN, established in 1945 to avoid future conflicts on the scale of World War I and World War II, is an international organization made up of 193 member states as of 2016. Written in 1945, the UN's charter (from www.un.org) contains four aims:

1. *to save succeeding generations from the scourge of war, which twice in our lifetime has brought untold sorrow to mankind,*
2. *to reaffirm faith in fundamental human rights, in the dignity and worth of the human person, in the equal rights of men and women and of nations large and small,*

3. *to establish conditions under which justice and respect for the obligations arising from treaties and other sources of international law can be maintained, and*
4. *to promote social progress and better standards of life in larger freedom.*

In order to strive toward achieving these four aims, four guidelines (also from www. un.org) were also established:

1. *to practice tolerance and live together in peace with one another as good neighbors,*
2. *to unite our strength to maintain international peace and security,*
3. *to ensure, by the acceptance of principles and the institution of methods, that armed force shall not be used, save in the common interest, and*
4. *to employ international machinery for the promotion of the economic and social advancement of all peoples.*

This basis of international cooperation provides a logical place to begin examining the development of the concepts of sustainability. At the end of the day, while each individual nation can work toward becoming more sustainable, pollutants that cause acid rain do not distinguish between borders, waste that accumulates in oceans does not follow international water law, and rivers that are dammed in one country may reduce flow in a second country downstream. These issues are complex. Therefore, taking a global perspective helps ensure that all countries are working toward similar common goals.

The first significant milestone for sustainability within the UN was the World Conservation Strategy, developed in 1980 (IUCN, 1980). In this document, sustainability was described through three goals:

1. *Maintain essential ecological processes and life support systems,*
2. *Preserve genetic diversity, and*
3. *Ensure sustainable utilization of species and ecosystems.*

These three goals mainly revolve around the concept of protecting the environment, with terms such as "ecological processes," "life support systems," "genetic diversity," "species," and "ecosystems." However, seven years later, in 1987, the UN released the Brundtland Commission Report, which is probably the most recognizable milestone in the UN's sustainability development (Brundtland, 1987). Within the Brundtland Commission, a theme was developed to qualify sustainability. The theme reads that sustainability "meets the needs of the present without compromising the ability of future generations to meet their own needs." This theme is independent of protecting the environment, but the concept of the environment is still woven into the fabric of the theme. It is interesting that this concept is almost identical to the Constitution of the Iroquois Nations, which states (in part): "Look and listen for the welfare of the whole people and have always in view not only the present but also the coming generations, even those whose faces are yet beneath the surface of the ground – the unborn of the future Nation."

SIDEBAR 01

To read more about the United Nations, visit their website at www.un.org, or scan the following QR code:

In 2002 the UN hosted a World Summit on Sustainable Development, which for the first time defined what are called the three pillars of sustainability: economics, environment, and social (UN, 2002). During this summit, a key theme was the commitment to "building a humane, equitable, and caring global society, cognizant of the need for human dignity for all" at the local, national, regional, and global levels. With this new solid foundation of the three pillars, future conferences and summits began formulating objectives and themes around sustainability. For example, the 2012 UN Conference on Sustainable Development put forth three objectives (UN, 2012):

1. *Poverty eradication,*
2. *Changing unsustainable and promoting sustainable patterns of consumption and production, and*
3. *Protecting and managing the natural resource base of economic and social developments.*

Through these three objectives, the three pillars of sustainability are clear, with **economics** clearly a part of the second and third objectives, **environment** also in the second and third objectives, and **society** spanning all three objectives.

To begin quantifying the objectives and themes surrounding sustainability which were developed over a decade of conferences and summits, world leaders came together in 2000 to adopt the United Nations Millennium Declaration. This declaration included a global partnership to accomplish eight goals by 2015. These eight goals, which are known as the Millennium Development Goals, are listed below (UN, 2007):

1. *Eradicate extreme poverty,*
2. *Achieve universal primary education,*
3. *Promote gender equality and empower women,*
4. *Reduce child mortality,*
5. *Improve maternal health,*

6. *Combat HIV/AIDS, malaria, and other diseases,*
7. *Ensure environmental sustainability, and*
8. *Develop a global partnership.*

The final report on the Millennium Development Goals released in 2015 stated the effort toward accomplishing these goals was the most successful anti-poverty movement in history (UN, 2015a). Significant progress was also made toward the other seven goals, such as decreasing child mortality rate by more than half and increasing primary school enrollment rate in developing regions to 91%.

To continue building on the success and momentum of the Millennium Development Goals, the UN developed a new resolution in 2015 which is more extensive and goes further by addressing root causes of the issues facing the world (UN, 2015b). This resolution, titled "Transforming Our World: the 2030 Agenda for Sustainable Development," broadened the scope of sustainability significantly with 17 goals. These goals are summarized in Table 1.1.

The UN has put Sustainable Development goals and Millennium Development Goals to practice through many channels, most noticeably through their Economic and Social Council and Secretariat. Through these "main bodies," the UN promotes and finances sustainable development, provides coordination and oversight, and builds partnerships. A Sustainable Development Summit was held in September 2019 to assess the progress being made toward achieving the 17 goals. It was concluded that while progress has been made in many areas, the progress is not advancing at the speed or scale which would be required to complete the goals by 2030. World leaders attending this summit called for a decade of action, pledging to mobilize financing, enhance national implementation, and strengthen institutions. At the center of the decade plan is the need to tackle growing poverty, empower women and girls, and address the climate emergency. As stated in the 2019 Sustainable Development Goals Report, the world is currently not on track to end poverty by 2030 (UN, 2019). Regarding empowerment of women and girls, of the 39% of the workforce women represent, only 27% of managerial positions are filled by women. Between 1998 and 2017, climate-related and geophysical disasters claimed an estimated 1.3 million lives. Significant efforts will be required at the global, local, and personal levels to see these goals accomplished in the coming decade.

While the United Nations has provided strong guidance on how to pursue sustainability and has shown leadership in the implementation of their policies, it is important to understand how we as Civil Engineers need to incorporate sustainability into our professional lives. Fortunately, the American Society of Civil Engineers has provided clear guidance on how to do this.

1.2   AMERICAN SOCIETY OF CIVIL ENGINEERS AND SUSTAINABILITY

The American Society of Civil Engineers, or the ASCE, was founded in 1852 and is the oldest engineering society in the United States. More than 150,000 people are members across 177 countries, which includes over 380 student chapters at

TABLE 1.1

2015 United Nations Resolution on Sustainability Goals

	Goals	Brief Description
1	No poverty	End poverty in all its forms everywhere
2	Zero hunger	End hunger, achieve food security and improved nutrition, and promote sustainable agriculture
3	Good health and well-being	Ensure healthy lives and promote well-being for all at all ages
4	Quality education	Ensure inclusive and equitable quality education and promote lifelong learning opportunities for all
5	Gender equality	Achieve gender equality and empower all women and girls
6	Clean water and sanitation	Ensure availability and sustainable management of water and sanitation for all
7	Affordable and clean energy	Ensure access to affordable, reliable, sustainable, and modern energy for all
8	Decent work and economic growth	Promote sustained, inclusive, and sustainable economic growth, full and productive employment, and decent work for all
9	Industry, innovation, and infrastructure	Build a resilient infrastructure, promote inclusive and sustainable industrialization, and foster innovation
10	Reduced inequalities	Reduce inequality within and among countries
11	Sustainable cities and communities	Make cities and human settlements inclusive, safe, resilient, and sustainable
12	Responsible consumption and production	Ensure sustainable consumption and production patterns
13	Climate action	Take urgent action to combat climate change and its impacts
14	Life below water	Conserve and sustainably use the oceans, seas, and marine resources for sustainable development
15	Life on land	Protect, restore, and promote sustainable use of terrestrial ecosystems, sustainably manage forests, combat desertification, and halt and reverse land degradation, and halt biodiversity loss
16	Peace, justice, and strong institutions	Promote peaceful and inclusive societies for sustainable development, provide access to justice for all, and build effective, accountable, and inclusive institutions at all levels
17	Partnership for the goals	Strengthen the means of implementation and revitalize the global partnership for sustainable development

universities in the United States alone. The ASCE has been active in promoting the importance of sustainability and has made sustainability a major focus area. The ASCE defines sustainability as:

A set of environmental, economic and social conditions in which all of society has the capacity and opportunity to maintain and improve its quality of life indefinitely without degrading the quantity, quality or availability of natural, economic, and social resources.

This definition clearly incorporates the three pillars of sustainability (economics, environment, and social) and draws on the latter UN work that also incorporates the concept of recognizing future generations. In addition to having a definition with several active initiatives, the ASCE has also incorporated sustainability into their Code of Ethics.

In 1914, the ASCE adopted a Code of Ethics, which is the "model for professional conduct" for the ASCE members (ASCE, 2006). Within this Code, the ASCE has four fundamental principles and seven fundamental canons. Sustainability is mentioned in the first principle: "using [engineer's] knowledge and skill for the enhancement of human welfare and the environment." This principle directly addresses two of the three pillars of sustainability, social and environment. In addition to the first principle in the ASCE's Code of Ethics, sustainability is mentioned in several of the seven canons. Canon 1 states that "engineers shall … strive to comply with the principles of sustainable development." Further discussion of Canon 1 indicates that if professional judgment is overruled, engineers should inform clients or employers of the possible consequences. In addition, engineers need to work for the advancement of safety, health, and well-being of their communities (social pillar) and the protection of the environment (environment pillar). Canon 3 continues the sustainability theme by asking engineers to endeavor to extend public knowledge of engineering and suitable development. This dedication by ASCE of incorporating sustainable principles into their Code of Ethics enforces the commitment of the civil engineering community in understanding and incorporating sustainable practices into the field. The question becomes at this point, how is this done?

In response to the UN's 2030 agenda and sustainable development goals, ASCE put forth a five-year roadmap to sustainable development (ASCE, 2019). This roadmap outlines a plan for transforming the civil engineering profession to increase the societal, environmental, and economic value of projects delivered, addressing such challenges as climate change, urbanization, and rapid technological development. Within the roadmap are four priorities for change:

1. *Sustainable Project Development: Doing the right project*
2. *Standards and Protocols: Doing the project right*
3. *Expand Technical Capacity: Transform the profession*
4. *Communicate and Advocate: Making the case*

Within each of the four priorities, there is a strategic goal as well as desired outcomes. Priority one requires engineers to begin to approach projects and engineering in a new way, shifting the focus from being primarily economic considerations and the end product to the potential needs addressed by a project. An example of an outcome for this priority is development of a new process for engineers to engage as a part of a multidisciplinary team in identifying and defining project needs before project approval and execution. This shifts the design methodology to utilizing desired benefits to define the project rather than simply relying on existing and potentially outdated standards. In many cases, this may lead to minimizing the need for new infrastructure or increasing the implementation of nature-based systems.

For example, the development of decentralized stormwater management protocols helps address the negative environmental impacts of existing standards which seek to collect and immediately transport stormwater to receiving waters. Past approaches achieve the goal of protecting crucial infrastructure by transporting stormwater away, but these approaches also inadvertently cause negative environmental effects on the receiving water bodies. Therefore, proposed improvements integrate retention and infiltration of stormwater to still protect crucial infrastructure but also provide positive aesthetic, recreational, and resource preservation impacts.

Priority two seeks to move standards and protocols from the traditional prescriptive nature to a more process-oriented and performance-based system. This can be accomplished by establishing, adopting, and implementing methodologies which produce sustainable infrastructure by meeting project owner requirements; significantly improving economic, environmental, and social performance of the project; accommodating a changing operating environment; incorporating risk and resiliency into design; and accounting for operations, maintenance, and end-of-life deposition. An example of a desired outcome for this priority is the development of new, higher-level standards for sustainable infrastructure which utilizes tools like ASCE's Envision rating system, which will be discussed in greater detail in Chapter 10.

Achieving priority three requires a transformation of civil engineering through expanding capacity to achieve visions and principles of sustainable development through training and professional development opportunities. Some outcomes associated with this priority include the development of a certificate program which demonstrates an engineer's understanding of sustainable development principles and implementation and a significant number of professionals pursuing this certificate.

Finally, the fourth priority centers on the communication and advocacy of this transformation. Communicating reasons for change with civil engineers, the public, and all stakeholders is essential for implementation. To see a shift in the profession where civil engineers and the public demand an environmentally, economically, and socially sustainable infrastructure is the end goal for this priority.

SIDEBAR 02

Writing a high-quality essay:

A well-written essay contains three components: an introduction, a body, and a conclusion. The introduction should gently guide the reader into the topic manner, starting with high-level discussion that sets the reader up to understand the purpose of the body content. The introduction should end with a topic sentence, which clearly states the main points of the body. This will allow the reader to be fully ready for the body of your essay, which usually contains two to three main points that you are trying to describe to the reader. These could be examples, arguments, or situations which form the skeleton of your essay. Within the body, you should support the points with ideas and facts that wrap the skeleton with muscle and create a clear picture of your discussion. After efficiently stating your main points, the conclusion is a recap of your introduction and body. No new information should be provided in the conclusion, and the reader should be able to obtain the gist of your essay from only reading the conclusion. This allows readers to gain a general idea of your essay, and if they are interested, they can read the entire document. Finally, engineering essays are generally written in third person. While they can be written in first person, take care as most readers are interested in the topic of the essay, and not the writer of the essay.

1.3   AS CIVIL ENGINEERS, HOW DO WE INCORPORATE SUSTAINABILITY?

This textbook seeks to provide civil engineering students with concepts and tools to allow them to begin driving toward these four priorities for change, incorporating sustainability as an essential component of the profession. After the introductory chapter, this book is divided into ten additional chapters, encompassing the three pillars of sustainability (Chapters 2–4), moving into applications of sustainability in the five primary areas of civil engineering (Chapters 5–9) as well as sustainable certification programs (Chapter 10), and finishing with a glimpse into tomorrow's sustainability (Chapter 11).

Chapter 2 will cover the **economic** pillar of sustainability. Tools to quantify economic measures of sustainability will be introduced, along with case studies and examples studying civil engineering projects that have incorporated economic aspects of sustainability. Specific tools include Life Cycle Cost Analysis, present/future/annual worth, rate of return, and benefit/cost ratio. Chapter 3 will explore the **environmental** pillar of sustainability. In this chapter, Life Cycle Assessment (LCA), Life Cycle Impact Assessment (LCIA), Product Category Rule (PCR), and Environmental Product Declaration (EPD) will be discussed. An example of a steel PCR will be reviewed, along with three EPDs revolving around different types of steel. Finally, the ecological footprint, water footprint, and planet boundary will be introduced. The third pillar of sustainability, **social**, will be covered in Chapter 4. The social pillar has not been quantified to the same depth as the economic and environmental pillars, but there are several tools available that could provide insight on potential social impacts of civil engineering. These tools include previously published articles in civil engineering journals, social media and civil engineering, the five documents produced by the UN, the Oxfam Doughnut, the Human Development Index (HDI), the Social Impact Assessment (SIA), and the British Academy's framework on purposeful businesses. In addition, emerging areas of social metrics will be introduced, such as those related to social media use in civil engineering. The concepts covered in the three sustainability pillars are summarized in Table 1.2.

While Chapters 2–4 provide foundational information as to quantitative and qualitative metrics to the three pillars of sustainability, Chapters 5–9 will delve deeply into specific applications of sustainability in the five primary areas of civil engineering. Chapter 5 is devoted to environmental applications of sustainability while Chapter 6 covers geotechnical applications of sustainability. Chapter 7 will explore structural applications of sustainability, and Chapter 8 will cover transportation applications. Finally, Chapter 9 will cover the fifth primary area of civil engineering,

TABLE 1.2

Key Concepts Covered in the Economic, Environment, and Social Pillars of Sustainability

Economic	Environment	Social
• Life Cycle Cost Analysis	• Life Cycle Assessment (LCA)	• Existing civil engineering examples
• Present/future/ annual worth	• Life Cycle Impact Assessment (LCIA)	• Social media and civil engineering
• Rate of return	• Product Category Rule (PCR)	• United Nations
• Benefit/cost ratio	• Environmental Product Declaration (EPD)	• Oxfam Doughnut
	• Ecological footprint	• Human Development Index (HDI)
	• Water footprint	• Social Impact Assessment (SIA)
	• Planet boundary	• British Academy's framework for purposeful businesses

construction management. Chapters 5–9 will cover a wide range of topics, from drinking water treatment to geothermal energy foundations, cross laminated timber to reclaimed asphalt pavement, and worker safety. Chapter 10 explores existing sustainable certification programs which provide another method for quantifying the performance of infrastructure as it relates to sustainability. In addition to introducing and applying sustainable concepts, these six chapters of civil engineering applications will provide extensive practice problems that utilize concepts taken directly from the NCEES Fundamentals of Engineering (FE) reference handbook. This not only will allow for additional practice in preparation for the FE exam, but will also provide insight into the broad scope of coverage on the FE exam. The content of Chapters 2–10 is graphically represented in Figure 1.1.

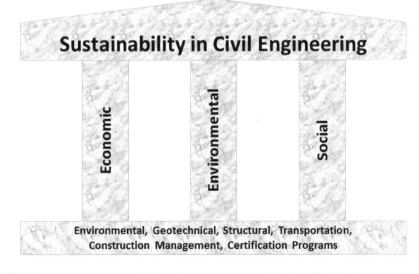

FIGURE 1.1 Components of civil engineering sustainability (credit: A Braham)

The book will finish with Chapter 11, which will give a survey of the future direction of sustainability in civil engineering. Sustainability is not a straightforward issue and the field itself is highly underdeveloped. In order to fully implement concepts from all three pillars, a paradigm shift will need to occur in industry and government. Engineers are typically strong in the STEM fields (science, technology, engineering, and math), but are less robust in the "softer skills" such as policy making and human development. This should be viewed, however, not as an obstacle but as an opportunity to continue identifying, building, and nurturing relationships across multiple disciplines in order to not only improve our world today but also tomorrow.

HOMEWORK PROBLEMS

For all answers in this chapter, use the format provided under the sidebar "Writing a high-quality essay."

1. The first paragraphs of this chapter discussed how ancient civilizations were either actively participating or providing governance in sustainable practices. Find a third ancient civilization that also incorporated sustainable practices and give three examples on how they did so.
2. The UN was established in 1945, but its aims have not changed since. Examine the four aims – do you think that all four are still important today? In your answer, provide three examples of why you think that they are still either important or not important.
3. Similar to question 2, the UN also has four guidelines that have not changed since 1945. Examine the four guidelines – do you think that all four are still important today? In your answer, provide three examples of why you think that they are still either important or not important.
4. Over the years, the United Nations hosted five conferences or summits (1980, 1987, 2002, 2012, and 2015) that directly revolved around sustainability. Of these five, which conference or summit do you think was most important in the development of sustainability on a global scale?
5. Of the 17 metrics developed during the 2015 UN Sustainable Development Summit, choose which metric you believe is most relevant and which metric you believe is least relevant. Provide two examples for each argument.
6. Two of the 17 metrics developed during the 2015 UN Sustainable Development Summit are directly related to Civil Engineering: water and infrastructure. Choose one of these two metrics and discuss three examples of sustainability in your chosen metric.
7. Eight goals were developed during the 2015 UN Sustainable Development Summit. Choose which goal you believe is most achievable by 2030 and which goal you believe is least achievable by 2030. Provide two examples for each argument.
8. After examining ASCE's definition of sustainability, do you feel it is complete? If so, justify with three discussion points. If not, provide three discussion points on how you think it could be improved.

9. There is no mention of cost or economics in the ASCE Code of Ethics. As one of the three pillars of sustainability, economics is an important component. Why do you think that economics was not mentioned in the Code, and do you think it should be included? Use three discussion points in your answer.

REFERENCES

ASCE. Code of Ethics. American Society of Civil Engineers, amended 2006.

ASCE. The Five-Year Roadmap to Sustainable Development. 2019.

Brundtland, G. H. Report of the World Commission on Environment and Development: Our Common Future. United Nations, 1987.

IUCN. World Conservation Strategy, Living Resource Conservation for Sustainable Development. International Union for Conservation of Nature and Natural Resources (IUCN), United Nations, 1980.

UN. Report of the World Summit on Sustainable Development. United Nations, Johannesburg, South Africa, 2002.

UN. Indicators of Sustainable Development: Guidelines and Methodologies. United Nations, Third Edition, 2007.

UN. Report of the United Nations Conference on Sustainable Development. United Nations, Rio de Janeiro, Brazil, 2012.

UN. Millenium Development Goals Report. United Nations, 2015a.

UN. Transforming our World: the 2030 Agenda for Sustainable Development. Resolution adopted by the General Assembly on 25 September 2015, Seventieth session, 2015b.

UN. The Sustainable Development Goals Report. United Nations, New York, NY, 2019.

2 Pillar
Economic Sustainability

We make a living by what we get, we make a life by what we give.

Winston Churchill

At the end of the day, private companies, public agencies, and all owners need to stay in business. This is often driven by financial considerations. If an organization is "in the red," it means that they are spending more money than they are making, and, in the long term, the organization will fail. However, many choices are only made by considering today's costs. So, if choice A costs less than choice B today, the organization will default to choice A. However, what if choice A costs less today, but will cost more over the 15-year design period versus cost B? Is it worth spending more money today to save money tomorrow? This is one of the key concepts of economics in sustainability, looking beyond today's cost and ensuring that, in the long term, the best economic decisions are being made. This concept is the cornerstone of the economic pillar of sustainability.

2.1 TRADITIONAL SUSTAINABLE ECONOMICS

Traditional economic considerations of sustainability revolve around three main points: local impact, material savings, and reuse. When considering economics and local impact, sustainable practices provide employment and stimulate local economy. By saving materials, that is, reusing existing materials, organizations can reduce upfront costs, reduce the transportation of materials, and reduce onsite waste. In addition, by utilizing fewer natural resources, future savings are gained in many areas, such as reducing the amount of material going to landfills. While these are all important concepts of reuse, they are limited in the fact that raw costs are not the only factor; what is more, maintenance and disposal costs may be quite different depending on the manufactured product or the engineering infrastructure. Finally, long-term performance is not taken into account, and in some applications, the longer-term performance may not even be known. These are certainly challenges while considering the economic perspectives of sustainability, but there are several concepts that can aid in more accurately capturing the full life span. These include life cycle cost analysis (traditional and probabilistic, Section 2.2), present/future/annual worth (Section 2.3), rate of return (Section 2.4), and benefit/cost ratio (Section 2.5).

2.2 LIFE CYCLE COST ANALYSIS

A life cycle cost analysis, or LCCA, is a very well-established method of quantifying long-term economic impacts. An LCCA takes into account both initial and discounted future costs in an attempt to identify the best value over the life of either a manufactured product or engineering infrastructure. A convenient feature of an LCCA, when comparing two different cost alternatives, is that common costs can cancel out and only costs that are different are considered. This highlights the importance of stating assumptions in the analysis, as different stakeholders in a project could make very different assumptions. Another key aspect of an LCCA is determining the analysis period. An analysis period can cover either a portion or the full life of a product or infrastructure and can even extend through the salvage of material at the end of life. Regardless of the analysis period that is chosen, however, care must be taken when stating assumptions and defining the analysis period to reduce confusion.

When considering the life cycle stages, both manufactured products and engineering infrastructure can be broken down into six stages. For a manufactured product, such as aggregate in Portland cement concrete (PCC), the first stage is material extraction. After extraction, the material is processed (stage two) in order to reach manufacturing (stage three). After manufacturing, the product is used (stage four), and finally, the product has an end of life (stage five). At this point, there is the possibility for the sixth stage, which is material reuse. For example, aggregates for pavements are either taken out of a quarry through a blasting operation (for manufactured aggregate), or can be dredged from a river bed (for natural aggregate). Generally, the aggregate needs to be processed by screening out non-aggregate material, or deleterious material, and washing excess clay or dust off the aggregate. Next, the aggregate is crushed in a jaw crusher, cone crusher, or impact crusher and then screened in order to achieve specific gradations for either a base course in a pavement structure or within the concrete layer. This stage is the manufacturing stage (stage three). Once the desired gradation is achieved, the aggregate is ready for use in the pavement structure. However, over time, the PCC weathers and ages under traffic loads and environmental conditions, and the aggregate reaches the end of its life. The sixth and final stage of a manufactured product is the use of the material in another project, for example, crushing the weathered and aged PCC roadway for use in a new application.

SIDEBAR 2.1 SALVAGE VALUE

Salvage value, in short, is the economic value of either a product or infrastructure at the end of the analysis period for the product or infrastructure. This can include the recycle, remanufacture, or reuse value of the product or infrastructure. If no specific data is available to calculate the salvage value, it is assumed to be zero. There are several methods for calculating or estimating the salvage value. For example, the Federal Highway Administration uses the following for pavements:

$$\text{Salvage value} = \left[1 - \left(\frac{\text{actual life alternative}}{\text{expected life of alternative}} \right) \right] \times \text{cost of alternative}$$

This simplified approach is acceptable by many because of the high level of uncertainty associated with service lives and costs for different pavement layer components, and the relatively small impact that salvage value has on life cycle cost comparisons. However, more complicated measurements for salvage value have been developed for pavements, including the following:

$$\text{Salvage value} = \left[\text{CLR} \times \frac{\text{remaining life of last resurfacing}}{\text{service life of last resurfacing}} \right] + \text{CRI}$$

where CLR is the cost of the last resurfacing and CRI is the cost of the lower asphalt layers remaining from the initial construction. This calculation of salvage value accounts for the in-place value of the pavement structure in addition to the remaining life of the last resurfacing.

The use of this simplified approach in estimating salvage value is justified by the fact that there are several uncertainties associated with the service lives and costs for the different pavement component layers, and the relatively small impact that salvage value actually has on life cycle comparisons.

A similar sequence of life cycle stages can be evaluated for engineering infrastructure. The first life cycle stage for engineering infrastructure is site development. The second is infrastructure manufacturing, which is followed by materials and product delivery (stage three). The fourth life cycle stage is infrastructure use, and the life cycle is complete at the end of the infrastructure's life and use in other applications. For example, during the construction of a bridge, the first stage is preparing the approaches and pylons for the support columns. The second stage would be the manufacture of the steel, concrete, and bridge deck material that will be utilized during construction. The third stage would be the actual delivery of the steel, concrete, and bridge deck material and assembly on site. Once the bridge construction is completed, the bridge is in use for the full life span, which eventually leads to the end of life. If the opportunity presents itself, the components of the old bridge can be reused (stage 6). For example, it is not uncommon for bridge girders to be taken from high-volume roads, sorted for functionality, and then used on lower-volume roads; this is a lower cost and a more sustainable option for small agencies such as cities and counties that can work with larger agencies, such as state and national agencies, in order to maximize the life use of each piece of infrastructure. A summary of the life cycle stages for both the manufactured product and the engineered infrastructure can be found in Table 2.1.

An advantage to defining projects with these six life cycle stages is the ability to clearly define three concepts: recycling, remanufacturing, and reuse. Going from the fifth step (end of life) to the second step (processing for materials, manufacturing for infrastructure) is recycling. Recycling is very common in materials such as steel and asphalt pavements, which are the two largest recycled materials by weight in the world. Remanufacturing occurs when moving from the fifth stage (end of life) to the third stage (manufacturing for materials, delivery and construction for

TABLE 2.1
Life Cycle Stages

Stage	Manufactured Product	Engineering Infrastructure
First	Material extraction	Site development
Second	Process material	Manufacturing of infrastructure
Third	Manufacturing of material	Infrastructure delivery and construction
Fourth	Product use	Infrastructure use
Fifth	End of life	End of life
Sixth	Reuse of material	Reuse of infrastructure

infrastructure). One example of remanufacturing is the utilization of existing facades in new buildings. When a historic building needs extensive renovation, a potential solution is to keep the facade of the building but completely replace the interior of the building. During this process, the facade needs a complete overhaul, which is a form of remanufacturing. The third concept of sustainability is reuse, which moves from the fifth stage (end of life) to use (the fourth stage). The previous example of reusing bridge girders from state roadways in country roadways is a good example of reuse. Within the three concepts of recycling, remanufacturing, and reuse, reuse has the highest level of sustainability. It requires the least amount of material transportation and processing. Conversely, recycling has the lowest level of sustainability, as materials generally need to be collected in a central location for processing and then redistributed to the field. Table 2.2 summarizes the concepts of recycling, remanufacturing, and reuse.

The traditional LCCA has six steps:

1. Establish alternative design strategies for the analysis period,
2. Determine performance periods and activity timing
3. Estimate agency and user costs,
4. Develop expenditure stream diagrams,
5. Compute net present value (NPV), and
6. Analyze results and reevaluate design strategies.

TABLE 2.2

Concepts of Sustainability, as Defined by the Six Life Cycle Stages, and Their Impact

From	To	Concept	Impact (Relative Sustainability)
Stage 5	Stage 2	Recycle	Lowest
Stage 5	Stage 3	Remanufacturing	Moderate
Stage 5	Stage 4	Reuse	Highest

The first step to LCCA, establish alternative design strategies for the analysis period, allows the engineer to decide what different options are worthy of being explored. While skyscrapers will probably never be constructed from aluminum (for both cost and material properties), the debate of steel versus PCC is always of interest, as is traditional steel designs such as moment frames versus diagrids. Therefore, understanding the options and then deciding the analysis period provide the foundation for the LCCA.

The second step to LCCA is determining the performance periods and activity timing. Typical performance periods include maintenance and rehabilitation schedules. Maintenance includes activities that have to be performed on a consistent schedule in order to allow the infrastructure to perform as designed, whereas rehabilitation occurs when more than maintenance is required but a full replacement is not necessary. A good example is a wood structure. If the wood is exposed, it needs to be sealed at least every 3–5 years in order to maintain integrity. Sealing is an example of maintenance. However, after several seals, often, individual pieces of wood must be replaced. It is anticipated that every 10–15 years, a portion of the structure will need to be replaced. However, the entire structure should not need to be replaced at 10–15 years if properly maintained.

The third step, and often the most difficult step for the LCCA, is estimating the agency and user costs. Here, not only the initial cost, but the maintenance cost, the rehabilitation cost, and the salvage value should all be quantified as well. Pavements are an excellent example of breaking out agency and user costs. For state agencies, examples of agency costs include materials, production, and construction for the initial cost. A popular comparison of two different pavements is asphalt concrete versus PCC. These have very different initial costs, which are highly dependent on geographic region and traffic level. However, user costs must also be considered. Not only is pavement construction a temporal inconvenience for drivers, it also costs the user money in lost productivity. In fact, the 2015 Urban Mobility Scorecard estimated that traffic congestion cost users $160 billion nationwide in 2014 (Schrank et al., 2015). This number is a combination of both fuel costs and lost time costs. While this is a combination of both construction congestion and standard congestion, it still shows the extreme scale of the impact of user delays on our economy.

The fourth step of the traditional LCCA is to develop an expenditure stream diagram. Continuing on with the pavement example, an LCCA analysis was done on Arkansas State Highway 98 on the cost of a 2-inch asphalt concrete overlay (Braham, 2016). The production, construction, and rehabilitation costs were estimated for a 50-year design period, with rehabilitation costs performed every 11 years. Note that there is no maintenance in this analysis, as it was assumed that maintenance costs were the same on all of the pavement types in the study. Figure 2.1 clearly shows the power of these expenditure stream diagrams; notice that it is very quickly obvious from the figure what the costs are and how they change over time.

The fifth step of the traditional LCCA is computing the NPV. The NPV takes all anticipated future costs and converts the costs to today's dollar value. When these converted future costs are added to the initial cost, a single number is created and

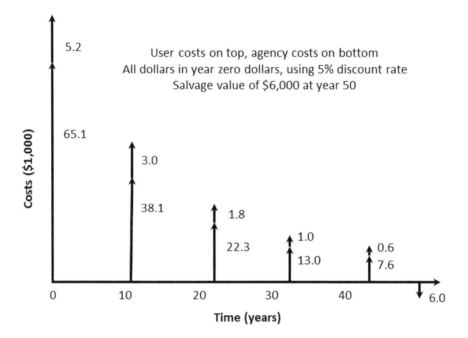

FIGURE 2.1 Expenditure stream diagram for a 2-inch overlay on AR98 for a 50-year design. (Credit: A. Braham.)

multiple alternatives can be compared directly. Equation 2.1 is used to compute the NPV.

$$NPV = \text{initial cost} + \sum_{0}^{t_n} \left(\frac{\text{maintenance cost}}{(1+r)^{t_n}} \right)$$

$$+ \sum_{0}^{t_n} \left(\frac{\text{rehabilitation cost}}{(1+r)^{t_n}} \right) - \left(\frac{\text{slavage cost}}{(1+r)^{t_n}} \right) \tag{2.1}$$

where
  t = time period analyzed (years)
  n = year of analysis
  r = discount rate (%)

This analysis shows the importance of determining performance periods and activity timing, as well as the importance of having available dollar values for initial maintenance, rehabilitation, and salvage costs. The combination of all of these four costs will provide a sound platform for making the final design strategy. The sixth and final step to the traditional LCCA analysis is comparing the NPV of different design strategies and reevaluating the strategies and assumptions to determine if any important details were overlooked.

At this point, it is important to note that the cost analysis done for the traditional LCCA has been deterministic. This simply means that single values were assumed for prices, both for initial costs and for future costs. However, this is an enormous assumption and not reflective of what actually happens in the real world, which has variability of the price of materials. A good example is the price of Portland cement.

According to the US Bureau of Labor Statistics, the price of Portland cement (using the 1982 price as 100 for an index-based analysis) bounced from 100 in 1982, to 199.3 in 2006, 209.7 in 2008, 187.8 in 2011, and 223.4 in 2015. This trend is shown graphically in Figure 2.2.

As shown in Figure 2.2, it is difficult to predict the price of Portland cement, so an alternative to a deterministic LCCA evaluation can be used. A popular alternative to a deterministic LCCA evaluation is a probabilistic LCCA analysis. The basic principles are the same as those of deterministic analysis. So, for example, the initial cost, the maintenance, the rehabilitation, and the salvage value would be determined. But instead of choosing a single number, a probabilistic analysis recognizes that there is a distribution of potential costs, especially when considering future costs. This analysis technique therefore includes uncertainty of future costs. This can be beneficial in three primary ways. First, the price of raw materials is rarely linear, and often single, unpredictable events (such as the 2008 housing crisis) can shift the global demand for the material. Therefore, while prices do generally go up over time, there are small levels of unpredictableness when looking at short timescales. Second, determining the proper discount rate is also very difficult. The discount rate is loosely related to bank interest rates, which do have a high level of variability, often as a function of the health of the economy. It is preferable to show a range of discount rates, from 2% to 6%, but if one value has to be chosen, a value commonly

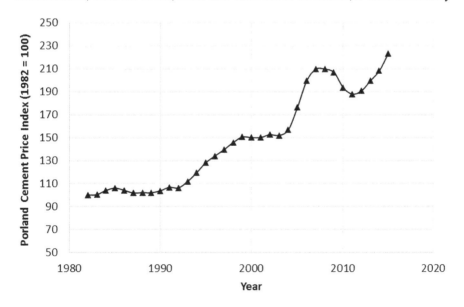

FIGURE 2.2   Unpredictable price of Portland cement. (Credit: A. Braham.)

chosen by engineers is 3%–5%. Finally, the third primary benefit of probabilistic analysis is that there are literally an infinite number of variables not only in the production and construction of civil engineering infrastructure, but also in how it will perform over time due to external factors. Therefore, if a historical rehabilitation schedule is used with new materials and equipment, there is a chance that rehabilitation may need to occur either more or less frequently (ideally, less with an increase in technology). Using a probabilistic analysis builds in these sources of uncertainty and variability, and allows for a more robust analysis.

Example Problem 2.1

In the state of Arkansas, a 2-mile stretch of an asphalt mixture four-lane road cost approximately $2,429,610 dollars to build in 2013. This cost included the unbound aggregate base course, the asphalt mixture base course, the asphalt mixture binder course, and the asphalt mixture surface course. Assuming a 50-year design life, a 4.0% discount rate, and the following rehabilitation, maintenance, and salvage value cost, what is the NPV of the roadway section? State any assumptions you needed to make.

- Maintenance (years 1–19): $3,300 per lane per mile
- Rehabilitation (year 20): $118,000 per lane per mile
- Maintenance (years 20–24): $1,100 per lane per mile
- Rehabilitation (year 25): $351,000 per lane per mile
- Maintenance (years 25–44): $2,900 per lane per mile
- Rehabilitation (year 45): $118,000 per lane per mile
- Maintenance (years 45–50): $1,100 per lane per mile
- Salvage value (year 50): $341,877 total

Assumptions are that maintenance will be performed in years of rehabilitation, and in year 50 of the design life, but will not occur in year 0. Equation 2.1 is used to calculate the NPV.

$$NPV = \text{initial cost} + \sum_0^{t_n}\left(\frac{\text{maintenance cost}}{(1+r)^{t_n}}\right) + \sum_0^{t_n}\left(\frac{\text{rehabilitation cost}}{(1+r)^{t_n}}\right) - \left(\frac{\text{salvage cost}}{(1+r)^{t_n}}\right)$$

The initial cost given = $2,429,610.

The maintenance calculation is split into four parts and totals $62,068.45 per lane per mile:

$$\sum_{n=1}^{19}\left(\frac{3300}{(1+0.04)^n}\right) = \$43,342.00$$

$$\sum_{n=20}^{24}\left(\frac{1100}{(1+0.04)^n}\right) = \$2324.33$$

$$\sum_{n=25}^{44}\left(\frac{2900}{(1+0.04)^n}\right) = \$15{,}375.40$$

$$\sum_{n=45}^{50}\left(\frac{1100}{(1+0.04)^n}\right) = \$1026.68$$

The rehabilitation calculation is split into three parts and totals $205,721.00 per lane per mile:

$$\text{Year}\,20 \rightarrow \left(\frac{118{,}000}{(1+0.04)^{20}}\right) = \$53{,}853.70$$

$$\text{Year}\,25 \rightarrow \left(\frac{351{,}000}{(1+0.04)^{25}}\right) = \$131{,}666.00$$

$$\text{Year}\,45 \rightarrow \left(\frac{118{,}000}{(1+0.04)^{45}}\right) = \$20{,}201.40$$

The salvage value calculation is given as $341,877.00. The maintenance costs and rehabilitation costs need to be multiplied by eight (as the prices above are per lane per mile, and it is a two-lane stretch of four-lane roadway). Therefore, the NPV is

$$\text{NPV} = \$2{,}429{,}610 + \left(\$62{,}068.45^* 8\right) + \left(\$205{,}721.07^* 8\right) - \left(\$341{,}877.00\right)$$

$$= \$4{,}230{,}250$$

2.3   PRESENT, FUTURE, AND ANNUAL WORTH

When analyzing economic alternatives, it is often convenient to present monetary amounts in different forms. For example, the LCCA discussion in Section 2.3 takes all future costs and brings them to one, single present cost. This has the benefit of essentially eliminating any concept of time from the equation and allowing for a direct comparison in today's dollars. However, in other situations, it may be more beneficial to know a single cost but at a future date. For example, if an agency is attempting to predict the funds available in 10 years for a highway rehabilitation, current funds may need to be converted into future dollars to, for example, account for inflation. Finally, a third common way of presenting money is in an annual form. When a company is attempting to compare the different energy costs associated with the heating and ventilation system in their building, for instance, it can be helpful to define costs on a yearly basis in order to better budget future costs. These concepts are summarized in Table 2.3.

TABLE 2.3

Three Forms of Representing Money

1. Single present cost	• Eliminates concept of time
	• Allows for direct comparison in today's dollars
2. Single future cost	• Current funds to future dollars
	• Predicts funds available and accounts for inflation
3. Annual future cost	• Defines future costs on a yearly basis
	• Beneficial for future budgeting

While these three different forms of presenting monetary costs have their pros and cons, one very useful technique for engineers is the ability to convert between the three forms. Conversions are generally done in three ways: future worth (F) to present worth (P), present worth to future worth, or present/future worth to annual worth (A). These abbreviations will be used extensively in the upcoming paragraphs.

When converting future worth to present worth (F → P), all future costs and revenues are converted into present dollars. This makes it easy to determine any potential economic advantage of one alternative over another. When presenting this concept, there are several common notations and equations. The official name is single-payment present worth, and the notation is (P/F, i, n), where P is the present value, F is the future value, i is the interest rate, and n is the number of years. Note that the interest rate in all of the present, future, and annual costs is in decimal form. The standard notation equation is $P = F(P/F, i, n)$, and the equation with factor formula is shown in Equation 2.2.

$$P = F\left[\dfrac{1}{(1+i)^n}\right] \tag{2.2}$$

In a similar fashion, the second useful conversion is converting present worth to future worth (P → F), where all present costs and revenues are converted into future dollars. This conversion is often used if an asset (such as a building) might be sold or traded after construction but before the expected end of life is reached. Using this technique, several alternatives' worth can be estimated at the time of either sale or disposal. The official name is single-payment compound amount, and the notation is (F/P, i, n). The standard notation equation is $F = P(F/P, i, n)$, and the equation with factor formula is shown in Equation 2.3.

$$F = P \times (1+i)^n \tag{2.3}$$

Finally, the annual worth (A), which is the equivalent uniform annual worth of all estimated costs and benefits during the life cycle of the alternative, is a useful tool for estimating yearly budgets. The annual worth can be calculated from either the

present worth or the future worth. When utilizing the present worth (P → A), the official name is capital recovery, and the notation is (A/P, i, n). The standard notation
equation is A=P(A/P, i, n), and the equation with the factor formula is shown in
Equation 2.4.

$$A = P\left[\frac{i(1+i)^n}{(1+i)^n - 1}\right]$$ (2.4)

Similarly, when the annual worth is known, and the present worth is desired (A →
P), the official name is uniform-series present worth, and the notation is (P/A, i, n).
The standard notation equation is P=A(P/A, i, n), and the equation with the factor
formula is shown in Equation 2.5.

$$P = A\left[\frac{(1+i)^n - 1}{i(1+i)^n}\right]$$ (2.5)

When utilizing the future worth (F → A), the official name is sinking fund, and the
notation is (A/F, i, n). The standard notation equation is A=F(A/F, i, n), and the
equation with the factor formula is shown in Equation 2.6.

$$A = F\left[\frac{i}{(1+i)^n - 1}\right]$$ (2.6)

Similarly, when the annual worth is known, and the future worth is desired (A → F),
the official name is uniform-series compounding amount, and the notation is (F/A, i,
n). The standard notation equation is F=A(F/A, i, n), and the equation with the factor
formula is shown in Equation 2.7.

$$F = A\left[\frac{(1+i)^n - 1}{i}\right]$$ (2.7)

Table 2.4 summarizes the discussed forms and equations of representing money.

Example Problem 2.2

The University of Arkansas has recently acquired a self-reacting frame that will
hold an actuator to perform dynamic testing on structural elements in the laboratory. Although the frame was generously donated by a local steel manufacturer,
the donation must be quantified for the development office. Assuming the present
value of the frame is $32,000, what is the single-payment compound amount and
the capital recovery? Use an analysis period of 10 years, and an interest rate of
3.5% for both calculations.
    To calculate the single-payment compound amount, use the following equation:

TABLE 2.4

Classifying Worth Conversion in Engineering Economics

How Conversion Is Performed	Official Name	Standard Notation Equation	Equation with Factor Form
Future worth to present worth	Single-payment present worth	$P = F(P/F, i, n)$	$P = F\left[\dfrac{1}{(1+i)^n}\right]$
Present worth to future worth	Single-payment compound amount	$F = P(F/P, i, n)$	$F = P \times (1+i)^n$
Present worth to annual worth	Capital recovery	$A = P(A/P, i, n)$	$A = P\left[\dfrac{i(1+i)^n}{(1+i)^n - 1}\right]$
Annual worth to present worth	Uniform-series present worth	$P = A(P/A, i, n)$	$P = A\left[\dfrac{(1+i)^n - 1}{i(1+i)^n}\right]$
Future worth to annual worth	Sinking fund	$A = F(A/F, i, n)$	$A = F\left[\dfrac{i}{(1+i)^n - 1}\right]$
Annual worth to future worth	Uniform-series compounding amount	$F = A(F/A, i, n)$	$F = A\left[\dfrac{(1+i)^n - 1}{i}\right]$

$$F = P \times (1+i)^n = \$32{,}000 \times (1+0.035)^{10} = \underline{\$45{,}139}$$

To calculate the capital recovery, use the following equation:

$$A = P\left[\frac{i(1+i)^n}{(1+i)^n - 1}\right] = \$32{,}000\left[\frac{0.035(1+0.035)^{10}}{(1+0.035)^{10} - 1}\right] = \$3{,}847.72$$

2.4   RATE OF RETURN

Care needs to be taken when calculating the rate of return, as the rate of return is based on the unrecovered balance and not on the initial balance. When an agency or firm borrows money, the interest rate is applied to the unpaid balance so that the total loan amount and the interest on the loan are completely paid with the last loan payment. The key is to determine the proper interest rate so that the final loan payment completely pays off both the total loan amount and the total interest amount. The rate of return is expressed as a percent per period (e.g., an interest rate per year).

To determine the rate of return, either the present worth form or the annual worth form can be used. If the present worth (or annual worth) of the costs is equated to the present worth of the incomes, the interest rate is called the root of the rate of return relation. If this root is equal to or larger than the minimum attractive rate of return, or MARR, then the alternative is economically feasible. If the root is less than the MARR, the alternative is not economically feasible.

The rate of return can be useful when issuing bonds, which is a common way to pay for large engineering infrastructure projects. Local and state agencies can often not afford to pay for mega projects, so they will issue bonds that investors can purchase. These bonds, over time, pay out with additional interest, which benefits both the agencies and investors. While these bonds may not have the highest rate of return, they are often more stable than other investment strategies and help round out an investor's portfolio.

Example Problem 2.3

The student union at the University of Arkansas has invested $225,000 to renovate the dining room. This is expected to save $9,500 per year for 20 years in maintenance cost of the room, and will save $300,000 at the end of 20 years in rehabilitation of the room. Find the MARR in order to establish that this alternative is economically feasible.

Convert all costs to year zero. Use trial and error to determine the increment when the ROR results go from positive to negative:

Use $-225,000 + 9,500(P/A, i, 20) + 300,000(P/F, i, 20)$:

$$i = 4\% \rightarrow -225,000 + 9,500\left[\frac{(1+0.04)^{20}-1}{0.04(1+0.04)^{20}}\right] + 300,000\left[\frac{1}{(1+0.04)^{20}}\right] = \$41,024$$

$$i = 5\% \rightarrow -225,000 + 9,500\left[\frac{(1+0.05)^{20}-1}{0.05(1+0.05)^{20}}\right] + 300,000\left[\frac{1}{(1+0.05)^{20}}\right] = \$6,458$$

$$i = 6\% \rightarrow -225,000 + 9,500\left[\frac{(1+0.06)^{20}-1}{0.06(1+0.06)^{20}}\right] + 300,000\left[\frac{1}{(1+0.06)^{20}}\right] = \$-22,494$$

Use linear interpolation to get the MARR

$$i = 5.00 + \frac{6458}{6458.3-(-22494)}(1.0) = 5.00 + 0.22 = \underline{5.22\%}$$

2.5   BENEFIT/COST RATIO

While the LCCA, present/future/annual worth, and rate of return are all useful tools in order to quantify alternatives economically, they are often used by corporations and businesses. Yet, a popular public service analysis technique is the benefit/cost ratio, or B/C. The B/C analysis technique was developed in response to the US

Congress Flood Control Act of 1936 and was designed to introduce a higher level of objectivity to public sector economics.

The first step is to convert all benefits and costs into a common equivalent monetary unit. Note that many assumptions and estimations go into this conversion, plus the analysis can change with different interest rates. The B/C can incorporate present worth (PW), annual worth (AW), and future worth (FW). The B/C is then calculated using one of the relationships shown in Equation 2.8.

$$B/C = \frac{\text{PW of benefits}}{\text{PW of costs}} = \frac{\text{AW of benefits}}{\text{AW of costs}} = \frac{\text{FW of benefits}}{\text{FW of costs}} \qquad (2.8)$$

When performing the benefit/cost ratio analysis, if the $B/C \geq 1$, the project is considered as economically acceptable. Conversely, if the $B/C < 1$, the project is not economically acceptable. Since the B/C is reported as a single number, clearly it produces an easy-to-understand result. As discussed in the previous sections of this chapter, there have been extensive tools provided to quantify economic alternative. However, if performed correctly, the B/C method will always select the same alternative as the present/future/annual worth and rate of return analysis techniques.

Example Problem 2.4

The Federal Aviation Administration (FAA) is awarding $12.5 million in grants to the Los Angeles World Airports in order to improve the lighting of the runways and taxiways at Los Angeles International Airport (LAX). These grants will extend over a 10-year period and will create an estimated savings of $1.35 million per year in energy costs and bulb replacement. The FAA uses a rate of return of 5% per year on all grant awards. The grants program will share FAA funding with ongoing activities, so an estimated $175,000 per year will be removed from other program funding. To make this program successful, a $435,000 per year operating cost will be incurred from the regular maintenance and operation budget. Use the B/C method to determine if the grants program is economically justified.

If $B/C < 1.0$, not economically justified
AW investment cost → $12,500,000 (A/P, 5%, 10) = $1,618,807 per year
AW of benefit → $1,350,000 per year
AW of disbenefit → $175,000 per year
AW of M&O cost → $435,000 per year
B/C = (AW of benefit − AW of disbenefit)/(AW of investment cost + AW of M&O cost) = ($1,350,000 − $175,000)/($1,618,807 + $435,000) = 0.57, not economically justified

2.6 SUMMARY OF THE ECONOMIC PILLAR

While there are many tools available to quantify the economic pillar of sustainability, the tool chosen is highly dependent on the information available, and the desired results of the analysis. The LCCA can compare multiple alternatives of either a manufactured product or engineering infrastructure, and determine from a monetary standpoint which alternative is more cost-effective. It is also necessary to convert dollar amounts to different forms. Here, present, future, and annual worth conversions can assist with issues such as inflation and budget planning. Other techniques, such as rate of return, exist to understand the consequence of choosing the correct interest rate when evaluating loan or bonding options. Finally, the B/C is convenient as it provides a single number comparison, which high-level politicians and administrators can quickly and easily process and understand. All of these tools have pros and cons, and it is up to the engineer to choose the most appropriate method for each unique situation.

HOMEWORK PROBLEMS

1. Using the data found in Example Problem 2.1, recalculate the NPV assuming a discount rate of 2.5% and 5.5%. How much does the NPV change compared to a discount rate of 4.0%?

2. Using the data found in Example Problem 2.1, recalculate the NPV assuming an initial cost of the pavement in 2007 dollars, of $2,177,437. In addition, recalculate the NPV assuming an initial cost of the pavement in 2010 dollars, of $2,432,059. How much does the NPV change compared to the initial cost of the pavement in 2013 dollars?

3. Using the format provided under Sidebar 1.2 "Writing a High-Quality Essay," compare the difference in using different discount rates (2.5%, 4.0%, 5.5%) and initial cost (2007, 2010, 2013 dollars). Which change has a great influence on the NPV? Use two discussion points in your answer.

4. A rural township in central Arkansas has recently replaced several septic tanks that have an anticipated life span of 24 years. Today, these septic tanks cost $24,000. However, they received a grant from the Environmental Protection Agency that matched the cost of the tanks today in order for the tanks to be replaced after their end of life. Assuming an interest rate of 7.5%, how much will a complete replacement of the septic tanks cost in 20 years?

5. Washington County, in northwest Arkansas, has plans to purchase $1.2 million worth of bridge girders from a bridge reconstruction on I-40 near Russellville. The purchase will occur in 20 years. These bridge girders will be used on small, low-volume bridges, and are an excellent example of the concept of reuse. Washington County, however, knows that they will not have $1.2 million in 20 years. Therefore, they have two options. They can either set aside a lump sum today, and over the 20 years, the amount they set aside will grow to $1.2 million, or, they can set aside a certain amount each year for 20 years and at the end of the 20 years, they will have $1.2 million. Calculate the total amount they would have to set aside just this year, and the yearly amount they would have to put aside in order to have $1.2 million after 20 years. Assume an interest rate of 6.3%.

6. A geotechnical engineer constructing a new levee in Arkansas has requested that $450,000 be spent now during construction on innovative geotextiles

to improve the stability of the slopes of the levee. This is expected to save $11,000 per year for 15 years in maintenance cost of the levee, and will save $625,000 at the end of 15 years in rehabilitation of the levee. Find the minimum rate of return in order to establish that this alternative is economically feasible.

7. The Arkansas State Highway and Transportation Department (AHTD) is investing $1,520,000 to be spent now for rehabilitation of I-30 near Arkadelphia, Arkansas. This is expected to save $66,000 per year for 30 years in maintenance cost of the roadway and will save $868,000 at the end of 30 years in rehabilitation of the roadway. Find the minimum rate of return in order to establish that this alternative is economically feasible.

8. The Beaver Water district has been granted a $5.3 million grant from the Environmental Protection Agency (EPA) to upgrade the sedimentation basin of the water treatment plant. This grant will extend over a 20-year period and will create an estimated savings of $735,000 per year in equipment and chemical savings. The EPA uses a rate of return of 4% per year on all grant awards. The grants program will share the EPA funding with the ongoing activities, so an estimated $65,000 per year will be removed from other program funding. To make this program successful, a $225,000 per year operating cost will be incurred from the regular maintenance and operation budget. Use the B/C method to determine if the grants program is economically justified.

9. A local consulting firm has been granted a $25,000 grant from the Bill and Linda Gates Foundation to upgrade the installed solar power panels on a local school. This grant will extend over a 5-year period and will create an estimated savings of $6,000 per year in energy costs. The Bill and Linda Gates Foundation uses a rate of return of 7% per year on all grant awards. The grants program will share Bill and Linda Gates funding with ongoing activities, so an estimated $2,500 per year will be removed from other program funding. To make this program successful, a $3,500 per year operating cost will be incurred from the regular maintenance and operation budget. Use the B/C method to determine if the grants program is economically justified.

REFERENCES

Braham, A. Comparing life-Cycle Cost Analysis of Full-Depth Reclamation Versus Traditional Pavement Maintenance and Rehabilitation Strategies. *Transportation Research Record: Journal of the Transportation Research Board.* 2016, 2573, 49–59.

Schrank, D., Eisele, B., Lomax, T., Bak, J. *2015 Urban Mobility Scorecard.* Texas A&M Transportation Institute and INRIX, College Station, TX, August, 2015.

3 Pillar
Environmental Sustainability

Perplexity is the beginning of knowledge.

Khalil Gibran

While a majority of businesses focus on economic repercussions of business decisions, almost all civil engineering infrastructure projects have an impact on the environment as well. This book will not debate concepts such as global warming, but it does recognize that construction of infrastructure and control of the environment often does produce emissions and waste.

In Chapter 2, topics such as LCCA, present/future/annual worth, rate of return, and benefit/cost ratio were reviewed. All of these metrics only consider the financial aspect of projects. An important question to ask, as well, however, through the lens of sustainability is: What sort of environmental impacts are there to a project? The environment is impacted during production, construction, use, and termination of roadways, wastewater treatment plants, dams, and buildings. Not only is energy used during all of these stages, but there are also wastes generated. Emissions are anything from carbon dioxide to volatile organic compounds (VOCs), and waste is anything from demolished structures to scalped raw material. Similar to the economic sustainability pillar section, there are several potential tools available to quantify the environmental impact of projects. These tools will help associate actual numbers with emissions and waste, which will then allow for an understanding of how designs could be changed to reduce such emissions and wastes. The first topic covered will be the structure of a life cycle assessment (LCA). The second topic will dive into details of the outputs of the LCA. After the comprehensive discussion of LCAs, examples of Product Category Rules (PCRs) and Environmental Product Declarations (EPDs) will be provided from the steel industry. Finally, this chapter will end with a brief discussion on other environmental qualifications and quantifications of sustainability, including Ecological Footprint (EF), water footprint, and planet boundary.

3.1 THE STRUCTURE OF A LIFE CYCLE ASSESSMENT

LCA is a technique to analyze and quantify the environmental impacts of a product, system, or process. In some ways, an LCA has a similar mission as the LCCA presented in Chapter 2, but instead of using dollars and cents, impact categories include energy use, resource use, emissions, toxicity, waste, and water use. LCAs are often used as a platform to support a Product Category Rule (PCR), which is used to develop Environmental Product Declarations (EPDs). PCRs and EPDs will be explored in detail later in this chapter. For the LCA, however, both ISO 14040:

"Environmental Management – Life Cycle Assessment – Principles and Framework" and ISO 14044: "Environmental Management – Life Cycle Assessment – Requirements and Guidelines" need to be followed. ISO stands for International Organization for Standardization and will be discussed in great detail in Chapter 9, Section 9.2, Sustainability Standards. For now, the focus will be on the structure of an LCA from ISO 14040 and ISO 14044.

Regardless of which area of Civil Engineering you work in, you will probably be able to find an LCA discussing a product, system, or process. While all of these LCAs are unique, the vast majority contain the elements required in ISO 14040 and ISO 14044:

1. Goal and scope definition of LCA
2. Life Cycle Inventory (LCI) analysis phase
3. Life Cycle Impact Assessment (LCIA) phase
4. Life cycle interpretation phase
5. Critical review of LCA
6. Reporting (including, but not limited to, limitations of the LCA; relationship between LCA phases; and conditions for use of value choices and optional elements)

These six elements allow for a comprehensive approach to evaluating the total environmental burden of a product, system, or process, by examining the inputs and outputs over the life cycle. The following will be a brief discussion of the six elements and will revolve around an example for flexible and rigid pavements from the Federal Highway Administration (Harvey *et al.*, 2016) and an example for flexible pavements from Michigan Tech University (Mukherjee, 2016). This brief discussion does not contain all of the required elements of an LCA but will provide enough of a foundation to provide the ability to read further on all of the details associated with an LCA.

The first element is the goal and scope of the LCA. Key aspects of the goal include the intended application, intended audience, and questions to be answered. For example, the goal of the flexible pavement LCA is "to support the Product Category Rule (PCR) for Asphalt Mixtures for the environmental product declaration (EPD) program hosted by the National Asphalt Pavement Association" (NAPA). This demonstrates one of the primary functions of an LCA, which is to provide the background necessary to first produce a PCR and then EPDs. A scope, on the other hand, can encompass many different components. In the case of the flexible pavement LCA, the most salient components are addressed:

• Definition of product: asphalt-based concrete (United Nations Standard Products and Services Code, UNSPSC: 30111509)
• Definition of functional unit: 1 US short ton of asphalt mixture
• Determination of life cycle stages and system boundaries (discussed below)

The life cycle stages and system boundaries are not straight forward. Starting with the life cycle stages, as discussed in Chapter 2, there are six life cycle stages of a manufactured product (such as flexible pavement): material extraction, processing material, manufacturing , product use, end of life, and reuse of material. Harvey *et al.* (2016) also indicated six life cycle stages, but they were slightly different: material production, pavement design, construction, use, maintenance/preservation, and end of life. However, there are other resources that divide the process into only four stages: material production, construction, use, and end of life (Harvey et al., 2011). Finally, Mukherjee indicated only three life cycle stages: production and construction, use, and end of life (Mukherjee, 2016). Therefore, it is important to know that there is no one way to capture life cycle stages. The key, however, is to clearly define the life cycle stages that will be used. In addition, the transportation of material between each life cycle stage is also often included in an LCA.

For example, in the flexible pavement LCA (Mukherjee, 2016), the life cycle assessment only covered what is called "cradle-to-gate," which comes under the production and construction life cycle stage. The cradle-to-gate for flexible pavement specifically includes raw material supply (aggregate, asphalt binder, other), transport of the raw material to the asphalt production facility, and the manufacturing of the asphalt concrete. Hence, "cradle" refers to extracting the material from the earth and "gate" refers to placing the asphalt concrete into the bed of a delivery truck. The flexible pavement LCA did not examine the construction of the pavement, use of the pavement, maintenance or rehabilitation of the pavement, nor the end of life of the pavement. Regardless of which stages are explored, each stage also has quite a bit of flexibility in the assessment, and this is where the system boundary comes in.

Examining just the cradle portion of the flexible pavement LCA, there are eight different materials within the system boundary: coarse aggregate, fine aggregate, liquid asphalt binder, Reclaimed Asphalt Pavement (RAP), Recycled Asphalt Shingles (RAS), polymers, additives (crumb rubber, antistrips, rejuvenators, fibers, etc.), and warm-mix technology and additives. Therefore, when determining the scope, it is important to ask which, if all, of these eight different materials need to be included. Another question is how wide the scope should be. For example, with RAP, is the milling of the material off an existing pavement, the transport back to the asphalt plant, the crushing and screening all included, or do you start at the processed RAP stockpile in the plant? There is no one answer to any of these questions of scope, but when you are reviewing an LCA or developing an LCA, these types of questions are critical to ask and answer.

The second element is the Life Cycle Inventory (LCI) analysis phase. The LCI compiles and quantifies all of the inputs and the outputs for a product through the established life cycle. The inputs are generally in the early life cycle stages and include energy input, raw material input, and ancillary inputs, whereas the outputs can be simplified to the product itself, the generation of wastes, and the generation of emissions. Table 3.1 describes some of the concepts that fall under these inputs and outputs.

The first input discussed is energy. Continuing with the flexible pavement LCA example (Mukherjee, 2016), a sampling of the energy needs and materials for just the plant operations of asphalt mixture include:

- Off road equipment to move aggregate
- Burner for drying aggregate

TABLE 3.1

Inputs and Outputs in a Life Cycle Inventory (LCI) Analysis

Inputs and Outputs	Descriptions
Input: for each life cycle stage	• Energy: renewable and nonrenewable energy sources, primary and secondary sources • Materials: renewable and nonrenewable material resources • Ancillary: inputs used to produce the product, but are not part of the product • Fresh water (potable)
Output: product	• Object of assessment and any co-products that are also marketable materials
Output: waste generation	• Hazardous and nonhazardous • Radioactive waste
Output: emission generation	• Emissions to air • Emissions to water • Emissions to soil

- Heating of liquid asphalt binder in the storage tank
- Movement of aggregate and liquid binder through the plant (pumps and drives)
- Asphalt mixture storage in silos
- Processing of RAP and RAS at plant site
- Additive processing and addition

Energy sources that need to be considered in this analysis include coal, nuclear, solar, wind, hydroelectric, biomass (for remote energy sources) and gasoline, diesel, natural gas, residual oil, geothermal, solar (for on-site energy sources). Some of these energy sources are renewable (solar, wind, etc.) and some nonrenewable (coal, gasoline, etc.). The materials for asphalt can also be split into renewable and nonrenewable materials resources. Renewable resources include RAP and RAS, whereas nonrenewable resources include aggregate, liquid binder, and additives. Continuing on with the inputs, ancillary inputs could include resources such as blasting and explosives used during quarrying operations. Finally, freshwater use (for dust control, for example) is important to capture, especially if it is potable. By using potable water for construction, drinking water is diverted away from human consumption.

The first output is the products. For flexible pavements, the product is asphalt concrete. However, a co-product that could be a marketable material is baghouse fines. These fines are the very smallest particles that are suspended in the drum during the mixing of the aggregate and asphalt binder. These very fine aggregate particles (smaller than P200) could be marketed and sold, thus providing a co-product during the manufacture of asphalt concrete. The second output are the wastes generated. According to Mukherjee (2016), there are no hazardous, nonhazardous, nor radioactive waste generated during the production of asphalt concrete. However, depending on the system boundary, a case could be made that there are sources of waste from the production of asphalt concrete. For example, over time, the oil that lubricates the pumps for the asphalt binder to move about the plan needs replacing. This oil may need to be disposed of. While there are many safe and proper mechanisms for disposing waste oil, it is one example of a potential waste from an asphalt concrete plant.

The final output is emissions generated. This is a broad topic. Potential emissions generated will be discussed in more detail in the next section. For the production of asphalt concrete, the only primary emission into the air, according to Mukherjee (2016), is stack emissions from the plant. This is a primary emission because it can be measured on-site and the actual data is used in the LCA. While there are no secondary emissions provided for the production of asphalt concrete, an example of secondary emissions from the transportation of the raw materials from the source to the production facility. These are secondary emissions because the actual emissions are not used for each truck that delivers the raw material, rather emissions are estimated from national databases that can estimate the amount of emissions per mile for various types of trucks. Therefore, for the LCA, only the type, number, and haul distances of each truck needs to be tracked that brings raw materials into the plant. The flexible pavement LCA utilized Franklin Associates 9999 Data for these truck emissions and also explored rail transport to the plants. For emissions to water and soil, an existing Life Cycle Inventory for Portland Cement (from the Portland Cement Association) was used to find secondary emissions for emissions to water and soil. This mix of primary and secondary sources is very common for LCAs, as they are extremely complex processes with many different inputs, outputs, and requirements. As mentioned, this discussion is not meant to be a comprehensive discussion of all of the topics that are necessary for an LCI, but it should provide the basic tools to begin a more in-depth exploration.

The third element is the Life Cycle Impact Assessment (LCIA) phase. The LCIA consists of the human or environmental damages and/or impacts from the LCI. The LCIA captures the significance and magnitude of all of the human or environmental impacts for a product through the life cycle of the product. For example, certain types of emissions (carbon dioxide, methane, nitrous oxide, etc.) from the LCI can be combined into different impact categories, such as global warming or ozone depletion. These impact categories often have some associated unit (such as Global Warming Potential with a unit of CO_2 equivalent and Ozone Depletion Potential with a unit of CFC-11 equivalent), which are called midpoint indicators. Finally, these midpoint indicators can be translated into an endpoint indicator, such as ecosystem loss due to sea-level rise or skin cancer due to UV radiation (for global warming and ozone depletion, respectively).

This process can be achieved in three steps, with the option of a fourth. The first step is the selection of impact categories. Then, each impact category is classified into the midpoint indicators, also referred to as category indicators. Finally, a characterization model is used to translate the LCI results into the impact category indicators. The optional fourth step would be to normalize, group, weight, or do some other sort of data quality analysis of the impact categories which will make the data more robust.

In the flexible pavement LCA example (Mukherjee, 2016), the emissions associated with electricity use, generator energy, plant burner energy, hot-oil heater energy, mobile equipment energy, aggregate, asphalt binder, and transportation are translated into five impact categories: 1) acidification, 2) eutrophication, 3) global warming potential, 4) ozone depletion, and 5) photochemical ozone formation. These five impact categories are described in more detail in the next section.

The fourth element is the life cycle interpretation phase. There are two different structures available for the life cycle interpretation phase, one through ISO 14044 (ISO, 2006) and the other through EN 15804 (CEN, 2013). EN stands for European standards maintained by the European Committee for Standardization. While both of these structures revolve around significant issues, assumptions, assessment of quality, limitations, and other similar concepts, Table 3.2 summarizes the differences in more detail.

For the life cycle interpretation phase for flexible pavement, Mukherjee (2016) begins with an analysis on how establishing a rigid impact percentage-based cut-off may be misleading. For example, if it is decided that a 1% contribution to overall global warming potential is a rigid cut-off, some inputs, that are in fact identical between two different asphalt mixtures, may be included in the first mixture but not in the second. This could occur if the first mix has shorter transportation distances for raw materials (thus increasing the percentage of the global warming potential). With the second mixture, and the longer transportation distances, the other factors would "appear" smaller in percentage and thus may be cut off.

Mukherjee (2016) continues with a sensitivity analysis that explores how sensitive the LCA is to reducing the asphalt binder content and the sensitivity for distance traveled for the delivery of the asphalt binder. When Reclaimed Asphalt Pavement (RAP) is added to a mix, it reduces the asphalt binder content. Therefore, mixtures with RAP would produce lower global warming potentials due to asphalt binder usage, but the study investigated if the decrease of asphalt binder provided a significant change. Finally, a sensitivity analysis was also performed for whether the asphalt binder was polymer-modified or neat. Overall, this LCA exercise provided several nice examples of the life cycle interpretation phase.

The fifth element is the critical review of LCA. An important component of the critical review is the audience of the LCA. For example, an LCA that is to be used for

TABLE 3.2
Structure of Life Cycle Interpretation Phase

ISO 14044	EN 15804
• Identify significant issues	• Results of study
• Completeness, sensitivity, consistency, and variability of the LCA	• Assumptions and limitations
• Conclusions, limitations, and recommendations	• Assessment of data quality
	• Disclosure of choices, judgments, and rationales

public communication requires a review by a panel of interested parties. However, if the LCA is simply for research purposes and to only advance LCA methodology, a critical review of an external expert is only recommended. If a PCR is developed, there are several steps to the critical review process. First, there should be a public consultation for the program, and an independent review should occur. Second, a third-party panel should verify the PCR. Finally, if an EPD is developed, it is recommended to perform an independent verification, but it is not required.

As mentioned earlier, the flexible pavement LCA was developed in order to produce a PCR and EPDs for NAPA. NAPA hired a university to write the LCA and PCR, and the PCR was approved by a three-member review panel and an independent third party. While the EPDs are developed through a private company, they are not verified. However, NAPA took the extra step to hire a consulting firm to review the online tool that generates the EPDs for NAPA member companies. It is also important to note that EPDs that have their roots in different LCAs or different PCRs cannot be compared directly, as the assumptions and data sources used would not be the same.

The sixth and final element is the reporting, which should include a summary of the study, an introduction, and any sort of additional requirements and guidance. The summary should include the motivation, background, methodology, and results of the LCA. The introduction contains a bit more detail on the motivation and background, whereas the additional requirements and guidance contain the required components of an LCA: goal and scope, inventory analysis, impact assessment, results, and interpretation. The full LCA report produced by Mukherjee (2016) can be found online, and contains all of these sections, along with some more detail.

Example Problem 3.1

In order to gain a different perspective on LCAs in Civil Engineering, Google the phrase "Life Cycle Assessment of Forest-Based Products: A Review" to explore wood's LCA (Sahoo et al., 2019). When examining this document, list the four areas that are aggregated from the LCI flows that comprise the LCIA phase.

The LCIA phase aggregates the LCI flows to explore four areas impacted: (1) resource depletion, (2) human health, (3) social health, and (4) ecosystems.

3.2   LCIA CATEGORIES

The results of an LCI come either in the form of compounds, such as carbon dioxide (CO_2) or methane (CH_4), or as combinations of emissions, such as greenhouse gas (GHG) or smog. The latter metrics have been developed to better understand the impact of humans on the earth. Table 3.3 summarizes the more common emissions and combinations of emissions used in an LCA.

The compounds listed in Table 3.3 are emitted during the production of materials (such as aggregate or Portland cement), the construction of civil engineering infrastructure, the use of infrastructure (vehicle emissions, water treatment), and end of life (milling pavement, building demolition). While the emissions are relatively

TABLE 3.3
Common Emissions Used in a Life Cycle Analysis

Compound	Common Name
CO	Carbon monoxide
CO_2	Carbon dioxide
SO_x	Sulfur oxides
N_2O	Nitrous oxide
NO_x	Nitrogen oxide
O_3	Ozone
CH_4	Methane
SF_4	Sulfur tetrafluoride
NH_4	Ammonia
CFCs	Chlorofluorocarbons
HCFCs	Hydrochlorofluorocarbons
CH_3I, CH_2BrCl, etc.	Halons
THC	Total hydrocarbons (including methane)
HCl	Hydrochloric Acid
HF	Hydroflouric Acid
VOC	Volatile organic compound (paintings, coatings, formaldehyde, etc.)
PM10 (PM2.5)	Particles measuring 10 μm (2.5 μm) or less
GHG	Greenhouse gas
Smog	Smog

straightforward and easy to quantify, additional metrics have been developed to better understand the effect of the combination of compounds. The two most common metrics are GHG and smog.

GHG is a combination of gases that contribute to the absorption and emission of radiation in the atmosphere. Without these gases in the atmosphere, the Earth would not be able to sustain life, as the temperature at the Earth's surface would approach zero degrees Fahrenheit. However, after the Industrial Revolution, the amount of GHG has increased significantly, with some estimates reaching 40% over pre-1750 levels. In a sense, this could potentially cause more of the Earth's heat to be trapped by the atmosphere, which is what is commonly referred to as global warming. The majority of the gases emitted that contribute to GHG are from the burning of fossil fuel and animal agriculture, and offsets have been reduced with deforestation.

The second metric, smog, was created as a by-product of the visual effect of pollution, commonly found over urban areas. Smog comes from two primary sources: emissions from coal power plants or emissions from transportation sources. When coal burns, it releases many emissions, such as carbon monoxide, carbon dioxide, sulfur dioxide, nitrous oxides, and PM10. While significant efforts have been made to clean the air as it escapes from coal power plants in the United States, some undesirable emissions still make it to the atmosphere. This problem is significantly higher in developing countries versus developed countries, as the cost of emission control at point sources is significantly high. The second main source of smog, transportation, often emits carbon monoxide, nitrous oxides, VOC, and various hydrocarbons (such as methane). High volumes of vehicles, especially slow-moving vehicles (such as those in traffic flow approaching jam density), exacerbate the emissions. What is more, once sources such as power plants or traffic emit emissions to create smog, a photochemical process can compound smog generation. When sunlight strikes nitrous oxide or VOC, it reacts with these emissions and generates even more smog. A continuous cycle is formed, where emissions that cause smog react with the sunlight to create more smog.

In addition to these single emissions and emission metrics that result from an LCI, several impact categories have been developed in order to better understand how individual emissions influence nature's natural air and water cycles. Figure 3.1 shows the different layers of the Earth's atmosphere. Four impact categories specifically will be explored here: ozone depletion, eutrophication, ocean acidification, and global warming potential (GWP).

The first quantification, ozone (O_3) depletion, attempts to capture the deterioration of ozone in the Earth's stratosphere. The general principle is that when atomic halogens (such as refrigerants, solvents, and propellants)are released into the atmosphere, they release chlorine molecules, which react with and destroy ozone (O_3) molecules. The reaction takes place as follows:

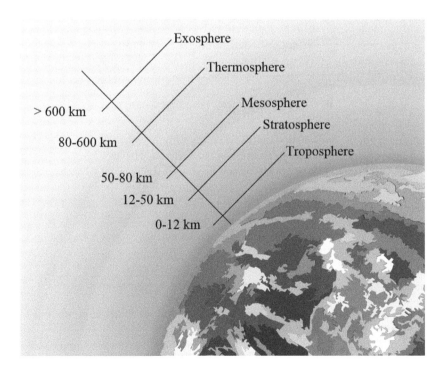

FIGURE 3.1 Layers of Earth's atmosphere. (Credit: William Crochot.)

$$Cl^- + O_3 \rightarrow ClO + O_2$$

$$ClO + O_3 \rightarrow Cl^- + 2O_2$$

The reaction process feeds on itself, creating, through the two reactions, more single chlorine atoms, which then react with more ozone. The single chlorine atoms react with ozone to create hypochlorite (ClO), which reacts with more ozone, creating more single chlorine atoms, and so on. The loss of ozone is most pronounced in the polar regions as the extreme cold promotes conditions that accelerate the two reactions above.

The second quantification, eutrophication, is the measurement of the addition of phosphates to the ecosystem. While eutrophication is a perfectly normal phenomenon in many cases, excessive phosphates can create highly toxic situations for native plants and fishes. Excessive phosphates can come from many sources, including detergents and fertilizers, and also from civil engineering sources such as wastewater. Point sources such as large-scale cattle or chicken farms, or sewage overflow devices can accelerate the detrimental process. The addition of these phosphates cause accelerated growth of plants and algae, especially in calm water sources such as ponds and lakes. These plants and algae consume oxygen, which can lead to hypoxia in the water. In general, aquatic life requires greater than 80% dissolved oxygen for a healthy environment. However, hypoxia occurs when dissolved oxygen is between 1% and 30%.

The third quantification, ocean acidification, quantifies a by-product of the natural ocean cycle where water attracts carbon dioxide. Like ozone depletion and eutrophication, ocean acidification occurs naturally, but human influence can accelerate the process, upsetting the natural equilibrium and causing significant changes in the ecosystem. In short, ocean acidification is the decrease in pH of the ocean from uptake of CO_2 in the atmosphere, and can be shown through the following reaction:

$$CO_2 + H_2O \rightarrow HCO_3^- + H^+ \leftrightarrow CO_3^{2-} + 2H^+$$

As can be seen, the carbon dioxide (CO_2) reacts with the water (H_2O) of the oceans, which then causes two potential reactions to create acid (H^+).

The fourth quantification, GWP, revolves around the concept of radiative force capacity. Radiative force capacity is the amount of energy (per unit area, per unit time) that is absorbed by GHG. A standard unit of radiative force capacity is W/m^2-kg/years. GWP is simply the ratio of the radiative force capacity of any substance over time of 1 kg by the radiative force capacity of carbon dioxide over time of 1 kg:

$$GWP = \frac{\text{radiative force capacity of any substance over time of 1 kg}}{\text{radiative force capacity of any } CO_2 \text{ over time of 1 kg}}$$

Based on this analysis, the GWP of carbon dioxide is one. Other common GWPs at different time periods are shown in Table 3.4.

This section has outlined many common emissions and compounds that can be used to quantify emissions from civil engineering applications, along with four quantifications that often occur naturally, but have also been accelerated with human

TABLE 3.4
Global Warming Potential of Several Common Emissions

Emissions		20 years	100 years
Methane	CH_4	86	34
Nitrogen oxide	N_2O	268	298
Sulfur tetrafluoride	SF_4	15,100	22,000

influence, especially after the Industrial Revolution. All of these metrics can be calculated during the production, construction, use, and end of life. The combination of the metrics and steps in the production of materials or infrastructure forms the complete LCA. With the LCA completed, the PCR and EPD can be addressed.

Example Problem 3.2

Google the document "Towards Sustainable Pavement Systems: A Reference Document" (Van Dam et al., 2015). One of the first links displayed should be from the Federal Highway Administration (FHWA), and is report number FHWA-HIF-15-02. Many governmental reports are available online for free download. Find the section that defines life cycle assessment. Define the acronym ISO and state what ISO's LCA definition is.

ISO stands for International Organization for Standardization, which, as its name implies, is an organization that represents multiple national standards organizations. ISO defined LCA as a process that

addresses the environmental aspects and potential environmental impacts (e.g., use of resources and the environmental consequences of releases) throughout a product's life cycle from raw material acquisition, through production, use, end-of-life treatment, recycling, and final disposal (i.e., cradle to grave).

This quote is found on pages 2–9 of the document (which is page 49 of the .pdf file).

3.3 PRODUCT CATEGORY RULE (PCR) AND ENVIRONMENTAL PRODUCT DECLARATION (EPD)

While an LCA can be seen as an exercise to gather information for an Environmental Product Declaration (EPD), the Product Category Rule (PCR) is a document that contains the rules to establish an EPD. The definition of a PCR, according to ISO 14025 is "Set of specific rules, requirements, and guidelines for developing Type III environmental declarations for one or more product categories." In general, a PCR should conform to ISO 21930 and will contain sections that cover these seven themes:

1. Product category definition and description
2. Goal and scope definition stages

3. Inventory analysis and additional environmental information
4. Impact category and inventory data categories
5. Materials and substances to be declared
6. Instructions for producing data, content, and format
7. Period of validity

When developing a PCR, a third-party review panel is required to verify that it is a complete document. Once the LCA and PCR are set, EPDs can be developed. An EPD is a document that is intended to communicate information about the life cycle environmental impact of a product. The document is intended to be transparent and comparable across multiple similar products, based on ISO 14025 and EN 15804. EPDs are for disclosure of information only and do not certify that any environmental performance standards are being met. According to ISO 14025, there are four objectives to an EPD:

1. To provide LCA-based information and additional information on the environmental aspects of products,
2. To assist purchasers and users to make informed comparisons between products,
3. To encourage improvement of environmental performance, and
4. To provide information for assessing the environmental impacts of products over their life cycle.

As mentioned earlier, an LCA is completed in order to produce a PCR and EPDs. However, EPDs from different LCAs and PCRs cannot be compared directly. Therefore, this discussion will take a single PCR for steel construction products (SCS, 2015) that leads to three EPDs (AISC, 2016a; AISC, 2016b; AISC, 2016c). The PCR will be discussed first and the discussion will follow the seven steps listed above.

The PCR for designated steel construction products, with SCS Global Services as the program operator, is an effort from multiple organizations that use steel and several private companies, including:

- American Institute of Steel Construction (AISC)
- American Iron and Steel Institute (AISI)
- Concrete Reinforcing Steel Institute (CRSI)
- Metal Building Manufacturers Association (MBMA)
- Steel Framing Alliance (SFA)
- ArcelorMittal
- Nucor Corporation.

As discussed earlier, the first topic covered in a PCR is the product category definition and description. The product category for the steel PCR includes fabricated structural steel, cold-formed steel sections, and concrete reinforcing steel. The second topic is goal and scope definition, which includes the stages covered. The goal of the PCR is to establish a set of consistent rules which will enable the creation of EPDs for fabricated structural steel, cold-formed steel sections, and concrete reinforcing steel. The scope definition, or the stages covered, include the production stage and "Module D." The production stage includes the extraction of material or recycling processes, transport to manufacturing facility, and manufacturing. These three items are often referred to as the "cradle-to-gate," as the cradle is the extraction/recycling of the steel and the gate is when the steel leaves the production facility. This steel PCR is somewhat unique as the Module D portion accounts for the reuse, recovery, and/or recycling potential of steel at the end of life. Since such a high percentage of steel is reused, recovered, or recycled, this PCR allows for those benefits to be accommodated in the EPDs. Finally, the functional unit of the analysis is one metric ton.

The inventory analysis begins with the activities during the cradle-to-gate analysis. For example, for the structural steel, this includes the recovery or extraction and processing of feedstock materials. Next, the furnace and related process for melting are analyzed, followed by casting and rolling into the final product. The PCR explicitly states that all upstream activities (such as fuel use and electricity generation) should be included in this stage. After the product is produced, the transport to the structural steel fabricator is accounted for, and the analysis finishes with the fabrication of structural steel elements. Another component of the inventory analysis is the co-product allocation. During the manufacture of steel, other value-added products are produced. For example, in a coke oven, co-products include carbon monoxide gas, coke, benzene, tar, toluene, xylene, and sulfur. These co-products, along with co-products developed in the blast furnace, basic oxygen furnace, and electric arc furnace, could distribute some of the emissions associated with the steel product. The final component of the inventory analysis is the calculation of the credit (or burden) associated with the recycling of steel at the end of life, and the incorporation of such steel for the next steel product at the cradle.

With the inventory analysis complete, the impact categories are defined. There are seven impact categories with associated parameters, many of which were discussed in Section 3.2, including categories such as ozone depletion and acidification. In addition, the LCIA method of obtaining the data is included. There are two different methods utilized in this PCR, TRACI (from the EPA) and CML Baseline (from the University of Leiden, Netherlands). These are tools that can be used to convert emissions that can be measured at a facility to an impact category. Finally, it is important to be clear on the required reporting units for each category, as there are often multiple units available to quantify impact categories. Table 3.5 summarizes the impact categories, parameters, method to obtain the data, and units.

In addition to the impact categories listed in Table 3.5, other waste categories include hazardous waste disposal, nonhazardous waste disposal, and radioactive waste disposal. Since the PCR specifically includes the Module D, four other output flows are necessary: components for re-use, materials for recycling, materials for energy recovery, and exported energy. Finally, there is a discussion and details provided on parameters that describe the resource use. This can include energy (renewable vs. nonrenewable), secondary materials, renewable and nonrenewable secondary fuels, and fresh water.

TABLE 3.5
Life Cycle Impact Assessment Results

Impact Category	Parameter	LCIA Method	Reporting Units
Global warming	Global warming potential (GWP)	TRACI	Metric ton CO_2 eq
Ozone depletion	Depletion potential of the stratospheric ozone layer (ODP)	TRACI	Metric ton CFC-11 eq
Acidification of land and water	Acidification potential of soil and water (AP)	TRACI	Metric ton SO_2 eq
Eutrophication	Eutrophication potential (EP)	TRACI	Metric ton N eq
Photochemical ozone creation	Formation potential of tropospheric ozone (POCP)	TRACI	Metric ton O_3 eq
Depletion of abiotic resources (elements)	Abiotic depletion potential (ADPelements) for non-fossil resources	CML Baseline	Metric ton antimony eq
Depletion of abiotic resources (fossil)	Abiotic depletion potential (ADPfossil fuels) for fossil resources	CML Baseline	MJ, net calorific value

The next portion of the PCR is the declaration of the materials and substances. For example, a full description of the use of each steel product is required, along with the product's materials (including coating). In addition to the materials and substances, there are instructions for producing the data and the content. This includes a review of the criteria for excluding any inputs or outputs (also known as cut-off rules) that are within the system boundary. Ideally, primary data should be used (this is data collected from actual facilities being used); however, secondary data can be used but must be clearly identified (for example, truck emissions are often calculated from national averages of the emissions for a type of truck per mile traveled). Finally, there is a discussion about the data quality requirements, which include precision, completeness, representativeness, consistency, reproducibility, sources, and uncertainty. Finally, the last item is the period of validity. PCRs are good for five years, and this PCR was approved on May 5, 2015, meaning it is no longer valid as of May 5, 2020. However, even if it is not currently valid, it still provides a general overview of what a PCR contains.

Once a PCR is generated, EPDs can be written. According to ISO 14025, there are four objectives to an EPD:

1. Provide LCA-based and other environmental information,
2. Assist purchasers and users to make informed comparisons between products,

3. Encourage improvement of environmental performance, and
4. Provide information for assessing environmental impacts of products over their life cycle.

The state of California, the Illinois Tollway, the Federal Highway Administration (FHWA), and other agencies are moving in the direction of incorporating or even requiring EPDs in various areas, including pavements. In fact, the California Department of Transportation (Caltrans) may require EPDs as of July 2021 for a minimum level of Global Warming Potential. However, this is a very fast-moving area, so it is recommended that you speak with whichever agency you are working with to understand the exact requirements. In addition to potentially being required by agencies, EPDs can count toward LEED certification, as will be discussed in Chapter 10. Therefore, there is incentive for different areas of Civil Engineering to generate EPDs for their products. EPDs can be developed to represent a "national-average" of a product, or can be developed for each individual contractor, facility, or even individual product.

The structure of EPDs is similar across different products. In general, EPDs begin with a description about the company that produces the product or an organization that represents the companies that produce the product. For example, the fabricated steel plate EPD (AISC, 2016a) provides a background for the American Institute of Steel Construction (AISC) and the National Steel Bridge Alliance, whereas the EPD for the fabricated hot-rolled structural sections only has a background on AISC (AISC, 2016b). The third EPD, for fabricated hollow structural sections (ASIC, 2016c), includes a background for AISC and the Steel Tube Institute. The next section is a summary of some of the logistical aspects of the EPD, which includes a discussion on the exclusions, accuracy of results, and comparability of the EPDs per ISO 14025, and then goes on to list details about the EPDs' stakeholders. All three of the EPDs for steel have UL Environment as the program operator and AISC as the declaration holder. The product is defined and briefly explained, along with the date of issue and period of validity. All three of the steel EPDs were issued in 2016 (two in March and one in December) and all three have a period of validity of five years. Finally, the PCR that each EPD was built from was reviewed by Industrial Ecology Consultants, who also verified the EPD and LCA.

With the logistical content completed, a product definition is provided, which includes a brief description, the participating members, the delivered product configuration, and application and codes of practice. For example, the hot-rolled structural steel sections (AISC, 2016b) state that more than 250 AISC members contributed data for the EPD development, and discuss how hot-rolled structural steel sections are often used for parallel flange sections, angles, channels, and tees in building, bridge, and industrial projects. Codes of practice listed for this product include Specification for Structural Steel Buildings (ANSI/AISC 360-10), Seismic Provisions for Structural Steel Buildings (ANSI/AISC 341-10), and Code of Standard Practice for Structural Steel Buildings (AISC 303-10).

The next section in each EPD is the Life Cycle Stages. All three EPDs cover the following life cycle stages:

1. Raw material extraction and processing (A1)
2. Transport to manufacturer (A2)
3. Fabrication (A3)

There is a brief description of each of these stages in each EPD. For example, in the fabricated steel plate EPD (AISC, 2016a), the raw material section discusses how the plates are manufactured entirely from structural steel and that all data was from a background dataset for North America. The transportation distances and modes (truck, rail, barge) were collected from each fabrication facility and for ancillary manufacturing materials (such as lubricants, gases, and welding electrodes). Finally, the manufacturing section provides a brief discussion on the manufacturing itself, galvanization, and includes the metal scrap generated during these processes. Finally, it explicitly states how environmental impacts associated with surface preparation and coating are not included in the EPD.

The next section in all three EPDs provides information on the requirements for the underlying life cycle assessment. For all three EPDs, the declared unit is one metric ton and the density of the steel is 7,800 kg/m^3. The system boundaries are the same as well, with an emphasis that the EPDs are cradle-to-gate, which means only the product stage is accounted for, but the construction stage, use stage, end-of-life stage, and benefits and loads are not accounted for. This concept, along with the shorthand notation for each stage (A–D) are shown in Figure 3.2.

Note how the benefits and loads section discussed earlier in the PCR are not included in these three EPDs. This is an example of how complicated the development of EPDs is, and care must be taken to ensure that proper data can be collected for each stage. The requirements for the underlying life cycle assessment continues with assumptions, allocations, and cut-off criteria. These sections are essentially identical between the three EPDs. Some assumptions include how fabrication data does not differentiate between the three different types of steel, and the only difference is the amount of time necessary for the manufacturing in the shop. In addition,

Product Stage			Construction Stage		Use Stage					End-of-Life Stage				Benefits and loads
A1	A2	A3	A4	A5	B1	B2	B3	B4	B5	C1	C2	C3	C4	D
Included in this study			Excluded from this study											
Raw materials supply	Transport	Manufacturing	Transport	Installation	Use	Maintenance	Repair	Replacement	Refurbishment	De-construction	Transport	Waste processing	Disposal	Reuse, recovery, recycling potential

FIGURE 3.2 Life cycle stages for all three steel EPDs

the electricity consumption was calculated by excluding companies that either mechanically or compressed air blasted their products, limiting the applicability of the EPD. Finally, the last assumption is that packing materials were excluded, along with capital goods for the production processes (machines, buildings, etc.). There are three cut-off criteria listed in each EPD:

- Mass: if flow is less than 1% of cumulative mass of the model it may be excluded
- Energy: if flow is less than 1% of cumulative energy of the model it may be excluded
- Environmental relevance: if the flow meets the above criteria exclusively, but is considered to have a significant environmental impact, it was included.

After the brief description about the company, the logistical summary, the product definition, the life cycle stages, and the requirements for the underlying life cycle assessment, the life cycle assessment results and analysis are provided. The first section of the LCA results and analysis is the use of energy and material resources. Table 3.6 summarizes the results of the three EPDs.

The second section of the LCA results and analysis is the life cycle impact assessment (LCIA). Table 3.7 summarizes the results of the three EPDs.

The last portion of the life cycle impact assessment shows a visualization of the primary energy demand for nonrenewable resources. The impact assessment categories include climate change, ozone depletion, ocean acidification, eutrophication, smog formation, and abiotic depletion (elements and fossil). For all of these impact assessment categories, the materials portion of the LCIA dominates the relative module impact, with well over 80% of each impact assessment category being attributed to the materials (A1) life cycle stage.

The next section of the EPD is the data quality assessment. This section discusses the temporal, geographical, and technological representativeness, along with

TABLE 3.6

Use of Energy and Material Resources (Results per Metric Ton)

	Steel plate (AISC, 2016a)	Hot-rolled structural (AISC, 2016b)	Hollow structural (AISC, 2016c)
Renewable primary energy resources (MJ)	489	509	731
Nonrenewable primary energy resources (MJ)	18,300	14,800	27,500
Secondary material (metric ton)	0.826	1.05	0.0683
Secondary fuels (MJ)	0		
Fresh water	Not included		

TABLE 3.7
Life Cycle Impact Assessment

	Steel plate (AISC, 2016a)	Hot-rolled structural (AISC, 2016b)	Hollow structural (AISC, 2016c)
Impact Assessment Method (TRACI 2.1)			
Global warming potential (GWP100) (metric ton CO_2 eq)	1.47	1.16	2.39
Depletion potential of the stratospheric ozone layer (ODP) (metric ton CFC-11 eq)	4.82E-08	2.25E-10	2.23E-08
Acidification potential of soil and water (AP) (metric ton SO_2 eq)	5.94E-03	5.94E-03	8.73E-03
Eutrophication potential (EP) (metric ton N eq)	2.16E-4	1.39E-04	4.38E-04
Formation potential of tropospheric ozone (POCP) (metric ton O_3 eq)	6.26E-02	3.12E-02	1.17E-07
Impact Assessment Method (CML 2001, version April 2013)			
Abiotic depletion potential (ADP-elements) (metric ton Sb eq)	1.42E-06	n/a	1.11E-07
Abiotic depletion potential (ADP-fossil) (MJ)	16,500	12,900	26,400
Other Environmental Information			
Hazardous waste disposed (metric ton)	n/a	2.65E-06	n/a
Non-hazardous waste disposed (metric ton)	n/a	1.52E-02	n/a
Radioactive waste disposed	6.61E-04	7.35E-04	4.30E-04
Components for reuse, materials for recycling, materials for energy recovery, exported energy	0		

precision. All three EPDs have text describing how the primary fabrication data were collected for 12 consecutive months during the 2013–2015 calendar years, and how the majority of the secondary data came from the 2016 GaBi databases (from Sphera). The data collected is intended to represent North America, and the actual primary data from manufacturers of steel plates, hot-rolled structural, and hollow structural steel was utilized. Secondary data used proxy data from North America. Finally, the precision of the primary data was considered high. The final section, contact information, simply provides contact information for the companies that produce the product and for the LCA practitioner, who in the case of these three EPDs was Thinkstep.

This discussion of EPDs revolved around three steel applications. However, it is quite easy to google other construction materials within Civil Engineering to find those material's EPDs. While each EPD is different, they all contain approximately the same information, but it could be formatted or presented in a slightly different manner. In addition, the development of EPDs has accelerated over the past ten years, so it is recommended to regularly see if there are new developments in your area for LCAs, PCRs, and EPDs. The intent of this content was to simply show you several examples that will allow you to start understanding the components of most LCAs, PCRs, and EPDs that have been completed and produced to date.

Example Problem 3.3

Vulcan Materials recently released an EPD for their 12 concrete aggregate products (by Googling "Vulcan Materials Environmental Product Declaration," the .pdf can be quickly found) (Vulcan, 2016). Vulcan decided to only analyze the "cradle-to-gate" life cycle of the aggregate, focusing only on the raw material supply, the transport, and the manufacturing. In this document, compare the impact results for the manufactured sand (MFG Sand) to the top sand (WCS).

When looking at Table 4 in the document, it is obvious that the manufactured sand has a higher impact than the top sand (or the natural sand). This is an interesting comparison because these are both very similar products as sands, but the manufactured sand requires more processing (extraction, crushing, and screening) versus the natural sand (extraction and screening). Essentially, the GWP, the acidification potential, the eutrophication potential, the ozone depletion potential, the use of net freshwater, the hazardous waste disposal, and the radioactive waste disposed are twice as much for the manufactured sand as for the natural sand. In addition, the photochemical ozone creation potential, the use of renewable and nonrenewable primary energy, and the nonhazardous waste disposal are greater for the manufactured sand than for the natural sand. Therefore, it is apparent that the crushing of aggregate is a significant portion of the emissions associated with aggregate production.

3.4 ALTERNATE ENVIRONMENTAL FRAMEWORKS: ECOLOGICAL FOOTPRINT, WATER FOOTPRINT, AND PLANET BOUNDARY

In addition to a Life Cycle Assessment, there are dozens of different frameworks available to quantify the impact that humans have on the environment. Three of the more known frameworks will be discussed here: ecological footprint, water footprint, and planet boundary.

The first framework, Ecological Footprint (EF), originated at the University of British Columbia in the early 1990s (Wackernagel, 1994). The concept is based on nature's capital, and the fact that certain needs are necessary for human life. These needs include healthy food, energy for mobility and heat, fresh air, clean water, fiber for paper, and clothing and shelter. The goal of the EF was to develop a scientifically sound calculation that could relate to clear policy objectives. In addition, it needed a clear interpretation, to be understandable to nonscientists, and to cover the functioning of a system as a whole. Finally, the metrics had to be based on parameters that are stable over long periods of time so that minor or local fluctuations would not compromise quantifications.

EF is based on taking specific economy or activity energy needs and converting that energy and matter to land and water needs. In short, this leads to a five-step calculation. First, the consumption of a city, region, state, or country is calculated and split into food, housing, transportation, consumer goods, and services. Second, land area of the analysis zone is appropriated into cropland, grazing, forest, fishing ground, carbon footprint, or built-up land. Cropland is the land available to produce food and fiber for human consumption, feed for livestock, oil crops, and rubber. Grazing is that land that can raise livestock for meat, dairy, hide, and wool products. Forest provides the land for lumber, pulp, timber products, and wood for fuel, while fishing grounds cover the primary production area required to support the fish and seafood caught. Forest is one category for providing wood products, the carbon footprint is the amount of forest land required to absorb CO_2 emissions. Finally, the last category is built-up land, which is the area of land covered by human infrastructure. Once the consumption and land use are identified, both resource and waste flow streams are calculated, which is the third step in the calculation. The fourth step is the construction of a consumption/land-use matrix. This matrix shows all categories of both consumption and land use, and it also indicates where there is not enough land for certain consumptions as well as which land is excess land. The deficiencies (either too much consumption or not enough land) give numbers greater than one while the excesses give numbers less than one. The fifth and final step sums all of the numbers and provides an estimate of EF for a region. These five steps are summarized in Table 3.8.

When considering EF from a country level, it is interesting to note that the highest EF countries are from the Middle East according to a 2010 report published by the Global Footprint Network (Ewing et al., 2010). This report states that the United Arab Emirates (UAE) and Qatar were producing EFs greater than 10.0 global hectares per person. This number states that if every person in the world was living the standard of living of the average UAE citizen living on UAE's resources, we would need over 10 Earths to sustain life. The next grouping down consists of fully developed Western countries, which required approximately 5–8

TABLE 3.8
Five-Step Calculation for Ecological Footprint (EF)

Step	Description of Each Step
1	Consumption of food, housing, transportation, consumer goods, and services determined
2	Land area appropriated into cropland, grazing, forest, fishing ground, carbon footprint, or built-up land
3	Resource and waste flow streams calculated
4	Construction of a consumption/land-use matrix
5	Sum all of the numbers provide an estimate of EF for a region

Earths to maintain their standard of living. The list continues down through second-world, developing, and third-world countries. According to the report, it is interesting to note that the countries requiring less than one Earth are quite diverse both geographically and socioeconomically, from the Democratic Republic of the Congo (population 63 million) to Bangladesh (population 158 million) to Puerto Rico (population 4 million).

There are, of course, some drawbacks to the EF concepts. First, the physical consumption-land conversion factor weights do not necessarily correspond to social weights. The analysis focuses 100% on the metrics at hand, but do not consider the social choices people have to make. These concepts are often ignored; they will be covered in depth in Chapter 4. Second, the EF does not distinguish between sustainable and unsustainable use of land, but only takes into account the land that is being consumed. Therefore, forests could be clear-cut or sustainably harvested, two processes to extract wood from nature, but the EF would treat these practices as the same. A third criticism is that in the EF model, there are many options to compensate for CO_2 emission and CO_2 assimilation, such as by forest, chemosynthesis, and autotrophs. However, the EF model only compensates for CO_2 emission and assimilation by forest, neglecting the other options. A fourth criticism is that there is a significant correlation between population density and resource endowment. As populations move away from rural living to urban living, the EF will increase significantly, especially as the analysis zone shrinks. This artificially inflates the EF of urban areas while perhaps underestimating the true EF of rural areas. The fifth and final criticism discussed here is that the EF is hard to use as a planning device. While it is noble to attempt to decrease the EF, there are few tangible concepts that agencies can focus on to begin the reduction, making it difficult to leverage. While no measure is truly perfect, these deficiencies have led to the development of other metrics, including the planet boundary.

The most current resource that provides a comprehensive overview of the EF was produced in 2019 (Wackernagel et al., 2019). This document provides the accounting procedures to quantify EF, EF data on individual nations, multiple examples, and strategies for redefining how our economies can succeed while being good stewards to the environment. Multiple case studies are also provided for users to examine various EF scenarios.

Example Problem 3.4

Google the document "Ecological Footprint Analysis San Francisco-Oakland-Fremont, CA" (Moore, 2011). What is the largest EF category within the "Transport" area of consumption, and how does it compare to the largest consuming EF area of consumption?

The largest EF category with the transport area of consumption is carbon. Overall, the carbon consumption is much high for transport versus "Food and nonalcoholic beverages." This is reasonable, since the food area of consumption relies heavily on cropland, whereas the transport area of consumption has carbon emissions associated with the activities.

The second framework, water footprint, was first proposed in by Hoekstra and Hung (2002). This document began the conversation about the water footprint by discussing international crop trade, and how "virtual water" flows between nations. In short, nations that have a low supply of water may want to import products that require a lot of water in production. This first study quantified the quantity of virtual water that flowed between nations from 1995 to 1999 for crops. The total water used by crops during this four-year period averaged 5,400 Gm^3/yr (7,063E+09 yd^3/yr). Overall, it found that 13% of the water (just under 175 Gm^3/yr, or 229E+09 yd^3/yr, average) used for crop production across the entire world is not used for domestic consumption but for export in virtual form. This document defined water footprint as "equal to the sum of the domestic water use and net virtual water import."

The next significant water footprint study was released by Chapagain and Hoekstra (2004). This study included not only international crop trade, but expanded into live animals, processed crop, livestock products, and industrial products from 1997 to 2001. These water flows were again defined in terms of virtual water, or the volume of water required to produce a commodity or service. This study updated the definition of water footprint to "the volume of water needed for the production of the goods and services consumed by the inhabitants of the country." The two-year stagger compared to the first water footprint study showed an increase of water use for crop production to an average of 6,390 Gm^3/yr (8,358E+09 yd^3/yr). Another interesting development in this study was the comparison of crop products' virtual water content versus livestock, where livestock are almost 5–10 times higher in water consumption versus crop products. Table 3.9 summarizes the virtual water content of various crop products and livestock. In addition, it was estimated that the agricultural products (crop production, live animals, processed crop, livestock products) constituted 80% of the world's virtual water flows and industrial products constituted 20%. Finally, the two largest factors that drove higher water footprints were high rates of evapotranspiration and high gross national income per capita.

A significant study on water footprints was produced by Hoekstra and Mekonnen (2012) that again significantly expanded the scope and detail of water footprints. In this study, consumptions of water were broken into three groups: rainwater (green),

---

TABLE 3.9

Average Virtual Water Content of Crop Products and Livestock from 1997 to 2001

Sector	Product	Virtual water content, m^3/metric ton, (yd^3/short ton)
Crop	Maize	900 (1,068)
	Wheat	1,300 (1,543)
	Rice	3,000 (3,560)
Livestock	Chicken	3,900 (4,628)
	Pork	4,900 (5,814)
	Beef	15,500 (18,392)

ground and surface water (blue), and polluted water (gray). The majority of all water goes to agricultural production, with crop production accounting for almost 90% of the water footprint, pasture accounting for just under 10%, and water supply in animal raising under 1%. When looking at just crop production, over 75% of the water footprint comes from rainwater, just over 10% comes from ground and surface water, and just under 10% comes from polluted water. Compared to agricultural production, less than 5% of the world's water footprint is industrial production and under 4% is domestic use. This breakdown can be viewed through the lens of internal water footprint and external water footprint as well. Seventy-eight percent of the world's water footprint does not cross national boundaries, while 22% is exported. These numbers are summarized in Figure 3.3

A comprehensive discussion of the goals and scope, the development, and assessment of water footprint can be found online in "The Water Footprint Assessment Manual" by Hoekstra et al. (2011).

The third framework, planet boundary, was first proposed in 2009 and is defined as a "safe operating space" for humanity (Rockström et al., 2009a). According to this theory, if human activities stay within the safe space, the Earth is able to absorb the human activities with no long-term harm to the environment; however, if human activities move outside of the safe space, the planet boundary theory states that long-term harm may occur to the environment. These spaces are associated with the Earth's biophysical subsystems and processes.

A major premise of planet boundary theory is that the environment has been unusually stable for the past 10,000 years, commonly referred to as the Holocene period. During this time, the Earth's temperatures, freshwater availability, and biogeochemical flows have all stayed within a narrow, stable range according to some research (Rockström et al., 2009b). However, with the beginning of the Anthropocene period, which started after the Industrial Revolution in the 1800s, human influence may have begun to damage the system that keeps Earth within the Holocene state

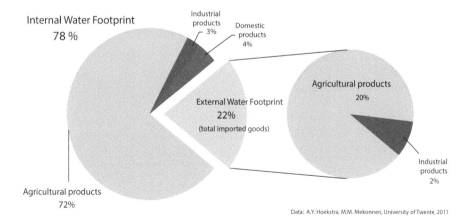

FIGURE 3.3 Global water footprint by sector (from Wiki, Sampa 2018 under attribution-share alike 4.0 International license)

according to planet boundary theory. This system was divided into nine subsystems, eight of which have been quantified. The nine subsystems have some overlap with both concepts learned in the LCA section and the EF section, but they also venture into new areas. The nine plant boundary subsystems are summarized and quantified in Table 3.10 for both 2009 and 2015 (Steffen et al., 2015), with a brief discussion following (Note: The nitrogen and phosphorus cycle are combined into one process called biogeochemical flows).

In Table 3.10, climate change is quantified by measuring the atmospheric carbon dioxide concentration, with units of parts per million by volume. Another quantification discussed for climate change was the change in radiative forcing. This radiative forcing is the same concept discussed earlier in Chapter 3 with the GWP, and was listed at 1.5 in 2009 and is set at 1.0 for the boundary. Additionally, the rate of biodiversity loss was measured by the extinction rate, and is the number of species per million species per year lost. The nitrogen cycle is the amount of N_2 removed from the atmosphere for human use in millions of tons per year, while the phosphorus cycle is the quantity of P flowing into the oceans per year, in millions of tons. The stratospheric ozone depletion is the concentration of ozone, using the Dobson unit, while the global freshwater use is the consumption of freshwater by humans per year, in kilometers cubed. The change in land use is simply the percentage of global land cover converted to cropland from the natural state of the land. The atmospheric aerosol loading is the particulate concentration in the atmosphere on a regional basis, and the chemical pollution, which includes emissions of everything from organic

TABLE 3.10
Planet Boundary Summary

Earth System Process Subsystems	Proposed Boundary	2009 Status	2015 Status	Preindustrial Value
Climate change (ppm CO_2)	350	387	396.5	280
Rate of biodiversity loss (extinctions per million species-years)	10	>100	100–1000	0.1–1
Biogeochemical flows: nitrogen cycle (million tons/year)	62–82	121	150	0
Biogeochemical flows: phosphorus cycle (million tons/year)	6.2–11.2	8.5–9.5	14–22	-1
Stratospheric ozone depletion (Dobson unit, DU)	276	283	As low as 200	290
Ocean acidification (carbonate ion concentration)	>2.75	2.90	2.89	3.44
Global freshwater use (km³/year)	4000	2600	2,600	415
Change in land use (percentage)	75	11.7	62	Low
Atmospheric aerosol loading (aerosol optical depth)	0.25–0.50	Not determined	0.30	Not determined
Chemical pollution	To be determined			

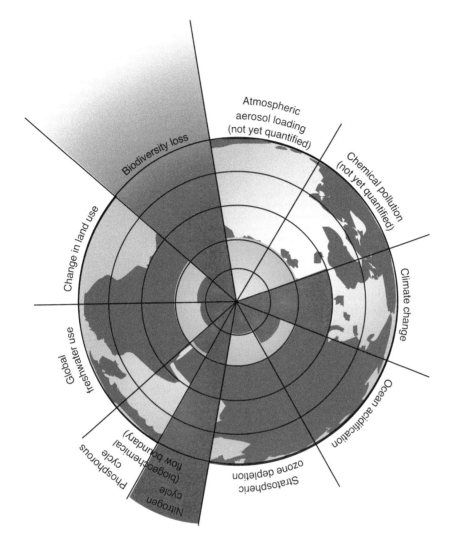

FIGURE 3.4   Planet boundary. (Credit: Azote Images/Stockholm Resilience Centre.)

pollutants, to plastics, to heavy metals, and nuclear waste, has not yet been quantified. Figure 3.4 shows a visual image of the processes.

Like the EF, planet boundary theory has some pros and cons, but it is another tool that can be potentially used to quantify the influence of civil engineering infrastructure on the environment.

Example Problem 3.5

On June 16, 2012, *The Economist* (a business magazine) published an article discussing planet boundaries titled "boundary conditions." The magazine article referenced a 2012 report by the Breakthrough Institute. The Breakthrough Institute report can be found by Googling "The Planetary Boundaries Hypothesis, A Review of the Evidence" and contains four discussion points. What are the four points?

The four discussion points highlighted by the Breakthrough Institute are: (1) non-threshold systems do not have boundaries, (2) the different means of mitigation and adaptation will have to be weighed against each other in pursuit of climate stability, based on their social, political, economic, and technical feasibility/desirability, (3) managing non-climate systems may not be appropriate for global regulation, but may require local- or regional-level regulation, and (4) the planet boundaries limits were established considering conditions during the Holocene (and it is difficult to prove that the Holocene period is the "optimal" period in Earth's history).

3.5   SUMMARY OF ENVIRONMENTAL PILLAR

Unlike the variety of established tools available for the economic pillar of sustainability, there is only one well-established tool to quantify the environmental aspects of sustainability, the LCA. While other concepts are available, such as EF, water footprint, and planet boundary, these have not been implemented across a wide range of civil engineering applications. However, there is promise in these developing fields, and they may become more salient in the future. Along with these three concepts, tools are being developed and implemented by industry, such as PCRs and EPDs, that outline the full environmental impact of products. Overall, the importance of the environmental impact of products and processes is becoming more pronounced, but there is still significant work to be done.

HOMEWORK PROBLEMS

1. Within the document "Life Cycle Assessment of Forest-Based Products: A Review," list the wood products and fuel/biomaterials that are manufactured from forest-based products. You should have 14 wood products and 9 fuels/biomaterials. Do not include any acronyms in your answer.
2. Within the document "Life Cycle Assessment of Forest-Based Products: A Review," what are the 17 modules listed for system boundaries for building products and building systems according to ISO 21930 and EN 15978?
3. Within the document "Life Cycle Assessment of Forest-Based Products: A Review," examine Table 5. Six buildings are listed in this table with four different system boundaries. Of the four different system boundaries, which system boundary do you think is the most appropriate for CLT mass timber buildings? In order to answer this question, use the format provided under Sidebar 1.2 "Writing a High-Quality Essay."
4. Within the document "Towards Sustainable Pavement Systems: A Reference Document," Chapter 3 discusses considerations to improve pavement sustainability. List the strategies for improving sustainability with regard to aggregate production. You should have three strategies.

5. Within the document "Towards Sustainable Pavement Systems: A Reference Document," Chapter 2 talks about general concepts of pavement sustainability. Within this chapter, six phases of a pavement's life cycle are defined. What are the six phases?

6. Within the document "Towards Sustainable Pavement Systems: A Reference Document," Chapter 10 talks in detail about assessing pavement sustainability. In Table 10.1, there are five impact categories listed for emissions. Which impact category do you think is the most important of the five? In order to answer this question, use the format provided under Sidebar 1.2 "Writing a High-Quality Essay."

7. Compare the EPD for Vulcan Material's concrete aggregate products to the description of the steel in the text. Using the format provided under Sidebar 1.2 "Writing a High-Quality Essay," discuss the two largest differences and the two largest similarities between the concrete aggregate and steel EPD.

8. Examine Table 4 in the EPD for Vulcan Material's concrete aggregate products. There are several impact categories that are listed as zero cradle-to-gate impact results. Why do you think that these values are zero, and how could Vulcan utilize some of these impact categories? Use the format provided under Sidebar 1.2 "Writing a High-Quality Essay" to formulate your answer.

9. Within the document "Ecological Footprint Analysis San Francisco-Oakland-Fremont, CA" the transport area of consumption is broken down into subcategories (Figure 9b). List the top three subcategories of the transport area. Were you expecting these three subcategories to be in the top three? Did you expect any of the other subcategories to be higher? Use the format provided under Sidebar 1.2 "Writing a High-Quality Essay" to discuss your answer.

10. Within the document "Ecological Footprint Analysis San Francisco-Oakland-Fremont, CA," Figure 1 shows a sampling of countries' EF. Why do you believe Qatar and UAE have such high EF values? Also, looking at the income level of the countries in Figure 1, can you develop a general relationship between income level and EF? Use the format provided under Sidebar 1.2 "Writing a High-Quality Essay" to discuss your answer.

11. Google "The Planetary Boundaries Hypothesis, A Review of the Evidence," and the .pdf article written by Nordhaus et al. (2012) should be one of the first links. Which of the nine Earth system process does the Breakthrough Institute believe can be successfully used on a global-scale threshold? Do you agree or disagree with their assessment? Use the format provided under Sidebar 1.2 "Writing a High-Quality Essay" to discuss your answer and include two primary discussion points.

12. Pick one of the nine Earth system processes within the Breakthrough Institute document and carefully review the institute's full analysis of the process. Identify one concept you agree with in their analysis and one concept you disagree with. Use the format provided under Sidebar 1.2 "Writing a High-Quality Essay" to formulate your answer.

REFERENCES

AISC. Fabricated Steel Plate. *Environmental Product Declaration*, Declaration number 4786979051.101.1, American Institute of Steel Construction, March 31, 2016, 2016a.

AISC. Fabricated Hot-Rolled Structural Steel Sections. *Environmental Product Declaration*, Declaration number 4786979051.102.1, American Institute of Steel Construction, March 31, 2016, 2016b.

AISC. Fabricated Hollow Structural Steel Sections. *Environmental Product Declaration*, Declaration number 4786979051.103.1, American Institute of Steel Construction, December 15, 2016, 2016c.

CEN. *Sustainability of Construction Works - Environmental Product Declarations – Core Rules for the Product Category of Construction Products*, European Standard EN 15804:2012 + A1, CEN-CENELEC Management Centre, Brussels, Belgium, 2013.

Chapagain, A., Hoekstra A. Water Footprints of Nations. *UNESCO-IHE*, Value of Water Research Report Series No. 16, Delft, Netherlands, 2004.

Ewing, B., Moore, D., Goldfinger, S., Oursler, A., Reed, A., Wackernagel, M. *The Ecological Footprint Atlas 2010*, Global Footprint Network, Oakland, CA, USA, 2010.

Harvey, J., Kendall, A., Lee, I., Santero, N., Van Dam, T., Wang, T. *Pavement Life-Cycle Assessment Workshop: Discussion Summary and Guidelines*. UCPRC-TM-2010-03. University of California, Pavement Research Center, Davis and Berkeley, CA, USA, 2011.

Harvey, J., Meijer, J., Ozer, H., Al-Qadi, I., Saboori, A., Kendall, A. *Pavement Life-Cycle Assessment Framework*, FHWA-HIF-16-014, Federal Highway Administration, Washington, DC, USA, 2016.

Hoekstra A., Hung P. Virtual Water Trade: A Quantification of Virtual Water Flows Between Nations in Relation to International Crop Trade. *UNESCO-IHE*, Value of Water Research Report Series No. 11, Delft, Netherlands, 2002.

Hoekstra, A., Chapagain, A., Aldaya, M. Mekonnen, M. The Water Footprint Assessment Manual: Setting the Global Standard, *Earthscan*, London, UK, 2011.

Hoekstra, A., Mekonnen, M. The Water Footprint of Humanity. *PNAS*, 2012, 109(9), 3232–3237.

ISO. *Environmental Management – Life Cycle Assessment - Requirements and Guidelines*, 1st Edition, ISO 14044. International Organization for Standardization, Geneva, Switzerland, 2006.

Moore, D. *Ecological Footprint Analysis San Francisco-Oakland-Fremont, CA*. Global Footprint Network, Oakland, CA, USA, 2011.

Mukherjee, A. *Life Cycle Assessment of Asphalt Mixtures in Support of an Environmental Product Declaration*, Michigan Technology University, Houghton, MI, USA, 2016.

Nordhaus, T., Shellenberger, M., Blomqvist, L. *The Planetary Boundaries Hypothesis, A Review of the Evidence*. The Breakthrough Institute, Oakland, CA, USA, June, 2012.

Rockström, J. et al. A Safe Operating Space for Humanity. *Nature*, 2009a, 461, 472–475, September.

Rockström, J. et al. Planet Boundaries: Exploring the Safe Operating Space for Humanity. *Ecology and Society*, 2009b, 14(2), 32.

Sahoo, K., Bergman, R., Alanya-Rosenbaum, S., Gu, H., Liang, S. Life Cycle Assessment of Forest-Based Products: A Review. *Sustainability*, 2019, 11, 4722.

SCS. North American Product Category Rule for Designated Steel Construction Products. *Product Category Rule*, Version 1.0. Approved May 5, 2016; valid through May 5, 2020. Program Operator: SCS Global Services, Emeryville, CA, 2015.

Steffen, W. et al. Sustainability. Planetary Boundaries: Guiding Human Development On a Changing Planet. *Science*, 2015, 347(6223), 1259855.

Van Dam, T., Harvey, J., Muench, S., Smith, K., Snyder, M., Al-Qadi, I., Ozer, H., Meijer, J., Ram, P., Roesler, J., Kendall, A. *Toward Sustainable Pavement Systems: A Reference Document.* Federal Highway Administration, FHWA-HIF-15-002, Washington, DC, USA, 2015.

Vulcan. Environmental Product Declaration for 12 Concrete Aggregate Products. Vulcan Materials Company, Western Division, 2016.

Wackernagel, M. Ecological Footprint and Appropriated Carrying Capacity: A Tool for Planning toward Sustainability. PhD dissertation, University of British Columbia, Vancouver, Canada, October, 1994.

Wackernagel, M. Beyers, B., Rout, K. *Ecological Footprint Managing Our Biocapacity Budget.* New Society Publishers, Gabriola Island, BC, Canada, 2019.

4 Pillar
Social Sustainability

Education is the most powerful weapon which you can use to change the world.

Nelson Mandela

The third pillar of sustainability, social, is the least quantified pillar in Civil Engineering compared to economic and environmental. The ongoing research in other disciplines, especially in the arts and sciences, has developed measurements for aspects of communities that have more success in addressing and solving problems. One such well-known community attribute is social capital. People are connected by social networks, and the exchange of trust and resources within those networks comprises measures of social capital. Community attachment is also recognized as another characteristic of engaged communities. The difficulty in measuring these well-known aspects of communities, however, lies partly in the differences between data sources, coverage, and availability. Much research has been performed with secondary data, often based on census data. This is because those publicly available datasets are available, affordable, and generally have widespread geographic coverage and large sample sizes. While many research projects collect primary data, primary data is more often limited to a relatively small population and/or geographic area as it is generally based on interviews or surveys. Primary data is expensive to collect and is also more difficult to use for generalizing because of limits in coverage, sample size, and/or comparability.

Metrics, to be effective indicators of a system, have four characteristics: relevancy, understandability, reliability, and accessibility. Relevance is key because the metric must provide information about the system one needs to know. Understandability is important so that even nonexperts can grasp the meaning of the metric. The metric must be trustable or reliable or the metric is of no use, and the data or information for the metric must be obtainable in a time frame suitable for decision making. The quandary for measuring societal sustainability and application to civil engineering comes in establishing effective metrics to answer three critical questions:

1. What level are we targeting for sustainability?
2. Who are we sustaining for?
3. Who gets to decide the answers to the first two questions?

In addition to these three questions and the difficulties with data mentioned earlier, other considerations are important as well. For example, if one area of society has a well-developed metric, does that influence other areas that do not have it? How do these metrics scale from a local or regional level upward to state, national, and international levels? Social metrics exhibit spatial heterogeneity, or unequal geographic distribution, which can further complicate scalar relationships. In general, many of the existing social metrics fall under four emerging areas: human well-being, access to resources, self-government, and civil society. These four emerging areas have provided much of the foundation of agencies and frameworks discussed in this chapter, including the UN, the Oxfam Doughnut, and the Human Development Index (HDI). There are many common themes throughout these three concepts. Another perspective of the social pillar of sustainability is through the Social Impact Assessments (SIA), which provide a framework for discussing societal impacts of Civil Engineering projects. This is similar to a Safety Data Sheet or an Environmental Product Declaration, but instead of focusing on safety or the environment, the SIA focuses on social issues. Finally, there has been significant movement in corporations in recognizing and addressing the social pillar of sustainability. While this will be discussed in more detail in Chapter 10 (Section 4), under the Environmental, Social, and Governance (ESG) section, an introductory discussion on the concept of the social purpose of corporations will be provided here.

Unlike the economic and environmental pillar, most existing social metrics have a short time horizon, and many are only available as cross-sectional data. Even those datasets that are longitudinal cover time-periods of a few decades, not 50 or 100 years. How will these metrics change over time frames appropriate for sustainability planning – in 10, 20, or even 50 years? This concept is explored in how groups of researchers are examining different perspectives of the social pillar of sustainability in Civil Engineering. One very specific and fast-growing area in Civil Engineering is social media, and how social media can be leveraged for the public's benefit, especially in the field of transportation engineering. While this chapter does not comprehensively answer all these revolving around quantifying the social pillar of sustainability in Civil Engineering, it does expose the reader to multiple tools and resources that can help identify potential paths forward.

4.1 EXISTING CIVIL ENGINEERING CONCEPTS

As discussed in this textbook's introduction, the American Society of Civil Engineers (ASCE) incorporates social aspects of sustainability into many portions of their Code of Ethics. In the fundamental principles, engineers are called to "use their knowledge and skill for the enhancement of human welfare" and to be "honest and impartial and serving with fidelity the public." This theme continues in the ASCE's canons, where engineers "shall hold paramount the safety, health and welfare of the public." The difficulty with following these charges in part comes with understanding the concepts behind specific terms. For example, what exactly is human welfare? What are the dimensions, attributes, and qualities of human welfare, and can these dimensions, attributes, and qualities be quantified?

There has been limited research performed specifically in the area of society and civil engineering. Yet, several groups within the civil engineering community have examined the issue. In 2007, for instance, Cheng et al. identified the need to measure sustainable accessibility in regional transport and land-use systems (Cheng et al.,

2007). They developed a model that utilized the average trip length and accessibility to jobs in an area and created a four-dimensional analysis that studied whether both parameters (trip length and access to jobs) were positive, negative, or a mix of the two. Accessibility was enhanced by either increasing the travel speed or bringing urban activities closer. Only car commuting was considered in this study. If a mix of trip length or access to jobs occurred, a better transportation system could be implemented or more jobs could be moved into an area. Similar to the gravity model, Cheng et al. also established that friction factors could be developed to represent other barriers associated with commuting from home to work. While the authors acknowledged the study was limited, they were confident it could provide a platform for further work. Another group that has examined the issue of society and civil engineering is Lucas et al., performing work in a similar area of regional transport and land use (Lucas et al., 2007). Lucas et al. focused on five areas of social sustainability, and they associated various engineering metrics with each one. The first area, poverty, was quantified examining total household expenditure on travel. The second area, accessibility, focused on weighted journey times to employment, education, health care, and food shops. The third area, safety, analyzed the number of child pedestrian casualties per 1,000 children in population. The fourth area, quality of life, captured the percentage of residents living within a 1-km or 15-minute "safe walk" to key destinations, including education, health care, leisure and cultural facilities, food shops, and the post office. The fifth and final area, housing, studied the lowest 10% value of house prices within the average local journey times to employment from the town center or other key centers of employment. Fields et al. sought to understand the relationship between transportation disadvantage and social exclusion specifically among lower-income older adults (2019). Study participants used a daily transportation diary app to share their transportation experiences as related to three domains of social exclusion: quality of life, participation, and resources. Based on responses, five primary themes relating transportation disadvantage and social exclusion emerged. These included constrained autonomy and flexibility, safety concerns, diminished emotional well-being, barriers to community engagement, and burdensome. This study captured qualitative data which contextualized lost opportunities and showed how economic disparities exacerbate the risk of transportation disadvantage. From these findings, valuable insight was gained for expanding conversations surrounding transportation planning, moving from a mobility focus to an equity focus. Additionally, this use of a digital platform for collecting holistic data related to transportation disadvantage potentially creates better opportunities for transportation planners, engineers, and social service providers to work together in addressing the needs of environmental justice populations.

As researchers have begun to qualify social aspects of sustainability in civil engineering, more focus has turned to identifying ways in which engineers can incorporate enhancing social equity and accessibility into transportation planning. Guthrie et al. sought to utilize general transit feed specification (GTFS) data to understand accessibility impacts of proposed transit projects in early planning stages to maximize social benefits associated with these large public investments (2017). GTFS is a data specification which allows public transit agencies to publish transit data in a more generally accessible format that can be utilized by a wide variety of software applications. Based on current GTFS data and proposed 2040 transit improvements for the Twin Cities region of Minneapolis-Saint Paul, Minnesota, Guthrie et al. developed a hypothetical transit network to explore impacts on access to job vacancies in historically disadvantaged areas. They found that quantifying accessibility during planning stages paints a comprehensive picture of the benefits offered by proposed improvements , allowing for more informed decisions to be made. This type of analysis also allows for more effective communication and debate surrounding planning decisions and provides a methodology for incorporating social sustainability into transit planning. Kuzio also conducted a research with a planning-focused approach, exploring how 20 different metropolitan planning organizations (MPO) prepare for emerging technologies and consider the implications on equity (2019). Results indicated that MPOs are making social equity an increasing priority; however, more consideration throughout the entire planning process is needed, including considering the implications of emerging technology on equity. By surveying approaches taken by MPOs across the country and of varying planning environments, this research sought to provide a starting point for planners and policymakers to understand how others are evolving policy and planning strategies to more effectively incorporate social sustainability.

Outside of accessibility and planning, additional work has considered social implications of transportation material selection. Alkins et al. examined the social benefits of in situ pavement recycling by exploring cold in-place recycling (CIR) (2008). This study examined the Ministry of Transportation Ontario's (MTO) promotion of using technology that reduces, recycles, and reuses, qualities deemed important for a society. CIR, which mills existing pavement in-place, stabilizes with a binding agent, and then places the material back onto the same roadway for an enhanced structural layer, fulfilling all three of these goals (reduce, recycle, reuse). They also discuss other social benefits of CIR, including improving safety. They found that safety is improved with the use of CIR by reducing traffic disruption and user inconvenience, reducing unsafe exposed edges and drop-offs (the milled pavement has a similar grade as the existing pavement after compaction), and expanding workers' ability to work through certain types of incremental weather since the material is placed at ambient temperatures. In addition, Alkins et al. point out that with all of the CIR work being performed in place, there is reduction in noise and disruption from traditional asphalt mixture production, transportation, and construction.

Another group of researchers have taken a unique approach to incorporate social sustainability into civil engineering education (Valdes-Vasquez and Klotz, 2011). In this work, both the traditional instructor to student teaching approach and the more innovative student to student teaching approach were utilized to convey social sustainability concepts. In both approaches, four dimensions of social sustainability were explored: community involvement, corporate social responsibility, safety through design, and social design. In addition to providing mapping and resources for both approaches, preliminary results were encouraging as initial feedback from students was positive for both the approaches. The authors concluded by recommending that other civil engineering programs also strive to incorporate social sustainability concepts into the curriculum.

Example Problem 4.1

List the potential social metrics that have been developed within existing civil engineering concepts.

The potential social metrics that have been developed within existing civil engineering concepts include time, accessibility, poverty, safety, housing location, and quality of life.

4.2 SOCIAL MEDIA AND CIVIL ENGINEERING

As social media has continued to play a more pivotal role in communication and culture, civil engineers have begun to explore ways in which it can be used to more effectively engage the public and therefore more completely address social aspects of sustainability. Research discussed in this section analyzed the use of social media to enhance disaster response, improve accessibility, and give users a voice in response and planning discussions.

Some researchers have begun to evaluate the use of social media in addressing crisis situations, such as extreme weather events. Social media provides real-time content, sentiment, and trends of public behavior which can prove to be invaluable in disaster preparation, warning, response, and recovery. Ukkusuri et al. explored the use of social media data to understand behavior during and after a crisis, specifically the tornado in Moore, Oklahoma, on May 20, 2013, to identify needs of individuals and groups affected (2014). The study focused on the content of social media posts on Twitter as well as the sentiment behind these posts. This information gave insight into disaster-related responses by different groups of users, issues of concern, changes in content during the event, and evolution of overall public attitude surrounding the event. Analysis included ranking the 10 most frequently used words and classifying tweets into 8 disaster-related categories. These words changed daily, and the trends were tracked. For example, "relief" was the tenth most frequently used word by news media on May 23, 2013, then jumped to the third most frequently used word on May 24 and 25. Ranking of words and tracking changes in most frequently used words indicated topics of most concern to each user group identified as well as the interests, needs, and concerns of these groups. The content analysis sorted tweets into categories of hazard location, warning, advice, damage or injury reports, help/aid/rescue, fundraising, support and consolation, as well as pet care. This research showed that the use of social media data can help emergency planners take necessary steps to improve disaster response and generally improve understanding of the behavior patterns of people during disaster events. From a civil engineering perspective, these steps could include more quickly identifying damage to infrastructure and the infrastructure of most interest to the population affected by the disaster. The ability to assess damage faster could allow engineers to return functioning to the infrastructure, such as a damaged roadway or bridge, more quickly. Sadri et al. studied patterns in social media usage during Hurricane Sandy in 2012 to enhance crisis communication procedures (2018). This analysis used a pattern recognition approach to identify word usage frequency and topics of interest during different phases of the storm as well as phase-independent topics. For instance, some topics which evolved included warning, response, and recovery; whereas, phase-independent topics were specific to location, time, media coverage, political leaders, and celebrity activities. This methodology can be used to identify real-time user needs in future crises, such as helping locate nearby shelters. Additionally, this type of analysis would allow emergency response officials to understand the evolution of concerns over time, which would help agencies prioritize responses and allocate resources. Table 4.1 summarizes frequent topics in each phase of the storm as well as examples of keywords used which are relevant to civil engineering. Many of these topics related to civil engineering deal with accessibility and transportation infrastructure. However, infrastructure damage could also be related to structural integrity of buildings or damage to water treatment facilities.

Looking more specifically at the application of transportation engineering, other researchers have explored the use of social media to communicate transit delays or traffic disruption. Chan and Schofer also conducted a case study around Hurricane Sandy, but they focused on the way transit systems in the New York region used Twitter to communicate with users during service shutdowns (2014). A graphical example of the damage caused to transportation infrastructure by Hurricane Sandy is shown in Figure 4.1. Three major area transit agencies were elected to disseminate information to users using Twitter as Hurricane Sandy approached and passed through the region. Consequently, these agencies saw a large increase in followers seeking to learn more about the transit shutdown, storm damage, and recovery actions. This illustrated the value of social media for transit agencies during severe weather disruptions to transmit information with more control of content and timing when compared to conventional communication channels. Additionally, social media allows for two-way communication as users can respond to posts, giving agencies user feedback and situational reports.

Pender et al. explored the use of social media to communicate unplanned transit disruptions which are not necessarily tied to a natural disaster event (2014). They conducted an international survey of 86 agencies on current practice of social media use. This study found social media to be useful in providing concise, real-time information to passengers, allowing them to make informed, proactive choices rather than reactive and suboptimal. Social media gives passengers a greater opportunity to take control of their transportation situation and elect alternative transportation rather than being confined to their typical route which has been disrupted. Douglass et al. also considered the utilization of social media to communicate transit disruptions and delays by seeking to understand transit users' needs and expectations for transit agency social media use (2018). An online questionnaire found that across age groups and travel purposes, the majority of people check social media before beginning their journey. This indicated social media has the potential to influence user travel plans before the journey, which could include changing routes, travel mode, and departure time. The questionnaire also found most users do not actively engage social media once the trip has begun, indicating potential to grow traveler engagement throughout the journey with the goal of improving user satisfaction. Outside of public transit, social media could also be used to support traffic management

TABLE 4.1

Summary of Storm Phases, Topics Discussed, and Relevant Key Words

Phase	Topic	Keywords Relevant to Civil Engineering
Warning	Storm prediction	Model, track, predict, impact, map, preparedness, plan
	Storm watch	
	Preparedness	
	User concern	
	Weather condition	
	Previous hurricane	
	Storm cause	
Recovery	Disaster relief	Airport, impact, fuel, recover, damage, cost
	Recovery	
	Damage/Aftermath	
Response	Gas/fuel	Fuel, shortage, flight, travel, airlines, subway, transit, bike, walk, bus, train, traffic light, bridge, tunnel, flood, evacuation, response, supply, truck
	Food/water	
	Power outage	
	Transportation	
	Local officials	
	Evacuation	
	Infrastructure	
	Death tolls	
	Hospital	
	Rescue	
	Fire	
	Flood	
	Trees/debris	
	Pets/Animals	
	Crime	
	Stock market	
	FEMA	
	First responder	
	Event cancellation	
	Work/school closure	

operations during social events or unplanned disruptive events. Wojtowicz and Wallace examined best practices for traffic operations among various agencies using social media to disseminate real-time, actionable information to motorists (2016). This analysis resulted in the following recommendations:

- Develop a social media policy, which is kept up to date, to define goals and objectives of social media use for traffic management as well as identify how the use of social media relates to the mission of the agency,

FIGURE 4.1   Damage to the road leading to Tom's Cove Visitor Center in Chincoteague, Virginia (Credit: Albert Herring, commons.wikimedia.org)

- Provide consistent levels of service over time to show users the agency's social media is a consistent and reliable source of information,
- Establish staff familiar with communication and the transportation system to implement and sustain the social media program,
- Use a consistent message structure which can be easily understood by the public,
- Incorporate the use of visual tools such as photographs whenever possible,
- Engage the public with relevant and useful information through personalized posts rather than computer-generated messages,
- Conduct outreach to make the public aware of social media tools available to them, and
- Collaborate with other agencies during nonroutine traffic events to deliver a consistent message to the public.

While the usage of social media discussed to this point has sought to improve responses to disruptions or disasters, an additional group of researchers has sought to evaluate the potential of social media from an even more proactive approach by incorporating its use into the transportation planning process. Evans-Cowley and Griffin considered social media for "microparticipation," which engages the public for purposes of maximizing information going into planning processes while minimizing the plan's development time and cost to the public (2012). Examining more than 49,000 posts on Twitter and other social networking sites, this research determined public engagement in the Austin Strategic Mobility Plan. Sentiment, extent of engagement, and impact on decision-making processes were all examined. With further refining, methods of data analysis used in this research could produce the ability for the public to influence decision making through future microparticipation efforts. This could be further extended to the use of location-aware social media, allowing users to tag planning issues such as traffic congestion, need for bike paths, and other transportation-related issues. As social media is incorporated into planning processes, it's important that the equity of access to government use of social media and effectiveness of informing and engaging the public through social media are measured. Camay et al. focused on the use of social media in the environmental review process of transportation planning (2012). They found social media to offer great potential for planners to enhance existing participation techniques; however, they also cautioned social media should not simply replace but be used to enhance traditional outreach methods. Enhancement could be realized by engaging a population who may not typically participate in more traditional formats, such as public hearings. However, if social media were to altogether replace public hearings, this could exclude a population that is not familiar or comfortable with this technology and that would prefer the traditional public hearings. Similar to recommendations made by Wojtowicz and Wallace, practitioners are recommended to devise a social media strategy with a defined goal, time frame, and target audience. This type of planning ensures communication preferences of various communities are considered in order to select the most appropriate tools and maximize the voice of the public. A sort of feedback loop should also be developed as a part of the social media strategy that measures the success of engagement by identifying qualitative measurements. Social media posts can also be utilized as data to allow planners to better understand travel behavior. Mjahed et al. explored how Yelp.com can serve as an information source for activity and trip planning (2017). In general, it was determined that Yelp reviews provide more information for car users, who can find access information for almost any business reviewed. Travel recommendations for walking and transit were generally related to walkability and transit supply. This study intended to serve as a launching pad for additional research to take a deeper dive into social media platforms to explore their role in trip planning and travel behavior. More complete understanding of factors influencing travel behavior will allow civil engineers to better address the needs and concerns of transportation users.

While this brief review has found that there have been several forays into concepts of social sustainability, there is not clear consensus on what social sustainability is, nor how it can be firmly measured with regard to civil engineering. In order to better develop tools, work done with the United Nations, Oxfam, the UN's HDI, and SIA is reviewed in order to gain a stronger foundation as to how other stakeholders are defining and exploring social impacts. In addition, a brief review of emerging areas will also be covered in order to provide stimulating thought moving forward in the field.

Example Problem 4.2

Describe how the applications of social media for use in civil engineering projects presented in this section could help address the social metrics introduced in

Section 4.1, including time, accessibility, poverty, safety, housing location, and quality of life.

Social media could provide a broader data source for calculating these metrics and provide a look at more real-time variation related to these social metrics, specifically as related to disaster response. Social media also allow for sentiment analyses which are concepts that are much more difficult to quantify but can provide an interesting dimension to social sustainability evaluations.

4.3 UNITED NATIONS, OXFAM DOUGHNUT, AND HUMAN DEVELOPMENT INDEX (HDI)

As discussed in Chapter 1, the UN has played a significant role in moving sustainability forward. This progress has not only occurred directly through various gatherings of the UN, but also through concepts that are borne out of the conclusions of the gatherings. To start, a brief review of the gatherings of the UN, and how these gatherings have discussed the social pillar of sustainability, will be provided. Next, two concepts that have been developed directly from the UN will be covered, the Oxfam Doughnut and the Human Development Index (HDI).

The first UN gathering that specifically introduced the social pillar of sustainably was the 2002 World Summit on Sustainable Development. This was the first time the UN adopted the three pillars of sustainability: economic, environmental, and social. Broad topics that fell under the social pillar included poverty eradication, changing consumption, and production patterns, human development, and the uneven distribution of the benefits and costs of globalization. These themes were continued and expanded on in 2007, when the UN released the third edition of indicators of sustainable development (UN, 2007). In this effort, the Commission on Sustainable Development (CSD) proposed 14 indicator themes and 44 subthemes, developed to emphasize the complexity of sustainable development and integration of the three pillars. Table 4.2 highlights the themes and subthemes that could potentially fall under the umbrella of the social pillar.

The themes in Table 4.2 were developed to be national in scope and rely on governments to develop metrics for local conditions. However, civil engineers can also begin incorporating these themes into their work in order to address the call from ASCE to incorporate sustainability and human welfare into our designs.

In 2012, the UN continued exploring the definition of social sustainability, reaffirming the need to eradicate poverty, change consumption and production patterns, and enforcing that people are at the center of sustainable development (UN, 2012). While several issues were discussed at length, specific actions toward societal advancement were included in the document, such as "promoting empowerment, removing barriers to opportunity, and enhancing productive capacity" for the poor and people in vulnerable situations, including youth and children. Finally, the most recent UN resolution proposed in 2015 revised sustainability metrics, again highlighting poverty, health, education, and consumption, but also focusing on women, inequality, and institutions (UN, 2015). The most discussed deliverable from this work are the 17 Sustainable Development Goals, which were discussed in Chapter 1.

TABLE 4.2
Themes and Subthemes from the UN Commission on Sustainable Development

Theme	Subtheme
Poverty	Income poverty, income inequality, living conditions
Governance	Corruption, crime
Health	Mortality, health care diversity, nutritional status, health status and risks
Education	Education level, literacy
Demographics	Population, tourism
Natural hazards	Vulnerability to natural hazard, disaster preparedness, and response
Atmosphere	Climate change, ozone layer depletion, air quality
Land	Land use and status, desertification, agriculture
Oceans, seas, coasts	Coastal zone, fisheries, marine environment
Freshwater	Water quantity, water quality
Consumption and production patterns	Material consumption, waste generation and management, transportation

Source: UN. Indicators of Sustainable Development: Guidelines and Methodologies. United Nations, 3rd Edition, 2007.

While few would argue that these are not important global issues, there has been little work (relative to the work done in the economic and environmental pillar) in tying these concepts directly to civil engineering and even less in developing quantifiable metrics that could potentially be incorporated into engineering design. One serious attempt at developing a comprehensive framework within which to develop social sustainability metrics is known as "the Oxfam Doughnut."

In 1942, Oxfam was founded in Oxford, England, in order to promote food relief through the Allied blockade for Greece citizens during World War II. Over the years, it has transitioned to an umbrella organization made of 19 partner organizations in 67 countries (Oxfam, 2019). The current charge of Oxfam is to "find practical, innovative ways for people to lift themselves out of poverty and thrive." As a part of this mission, they began a campaign that focused in part on developing a broad definition of prosperity in a resource-constrained world. A primary deliverable from this work was the Oxfam Doughnut, which is seen in Figure 4.2.

The Oxfam Doughnut is constructed from concepts discussed earlier in Chapters 1 and 3. Overall, the goal of the doughnut is to ensure a "safe and just space for humanity" that is inclusive of sustainable economic development. The heart of the Oxfam Doughnut focuses on societal issues and is directly related to concepts developed by the UN. The societal issues focus on critical human deprivations and outline societal foundations, including food, water, income, education, resilience, voice, jobs, energy, social equity, gender equality, and health. The outside of the doughnut

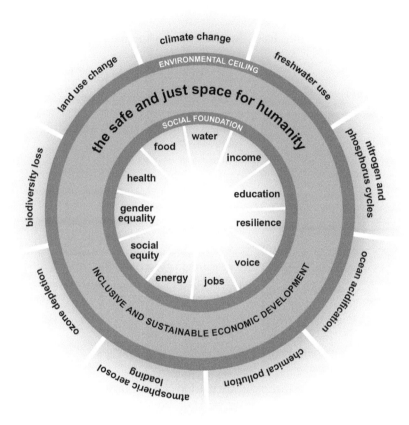

FIGURE 4.2 The Oxfam Doughnut. (Credit: Raworth, K. *A Safe and Just Space for Humanity, Can We Live within the Doughnut?* Oxfam International, February, 2012.)

focuses on environmental concerns and is directly related to concepts developed in the planet boundary theory, covered extensively in Chapter 3 and developed by Rockström et al. (2009). These concepts are the environmental ceiling of critical natural resources and planetary processes, including climate change, freshwater use, nitrogen/phosphorus cycles, ocean acidification, chemical pollution, atmospheric aerosol loading, ozone depletion, biodiversity loss, and land-use change. While the inside and the outside of the Oxfam Doughnut are taken from existing resources, the interactions discussed could lead to better quantifications of social metrics in civil engineering.

One such interaction is the relationship between environmental stress and poverty. Poverty has again and again been shown to be a critically important social issue facing the world today. However, people who are in poverty often have the fewest tools available to deal with environmental stress, whether it be flooding, severe drought, or other extreme weather events. A second interaction is that policies aiming for

FIGURE 4.3   The Biofuel logo (Note: this is under public domain on wiki)

sustainability in one area can exacerbate poverty in another area. A good example is that of biofuels, which have been represented by the biofuel logo seen in Figure 4.3.

In theory, using a renewable resource that is plant based to fuel vehicles (as opposed to petroleum, which is nonrenewable) is a brilliant idea. However, this approach has a significant disadvantage in that the resources to produce biofuels are the same ones that are used to feed people. Biofuel production may negatively impact hungry people in two ways. First, biofuel production diverts food production, making less food available for those who are hungry. Second, increases in the price of food are linked to biofuel production, as in theory, there is less net food available for human consumption and competition for agricultural resources, so prices rise. Increased cost of food results in poor people being less able to afford the food that is produced. This does not mean that biofuels should stop being used, but it does mean that when new ideas and technologies are being implemented, care should be taken to evaluate potential repercussions and unintended consequences, especially in areas not generally associated with the problem at hand. The goal is to develop solutions and policies that promote both poverty eradication and sustainability.

One such success story includes insulating homes. By properly insulating homes, energy bills decrease, winter deaths decrease, and inhabitant productivity increases. Insulation, if done correctly, can easily be sustainable, by using either recycled or prefabricated materials and easy-to-ship designs that can be implemented at low costs with minimal tools. Successful insulation of homes achieves success in all three pillars of sustainability: economic, environmental, and social. Figure 4.4 shows one well-insulated house (the purple house) in a row of less insulated houses (yellow, orange, and red houses), clearly showing that strong insulation reduces the amount of heat that is lost through walls during a cool day. The purple colors indicate a cooler than ambient surface (approximately 1.0°C less), whereas the red surfaces indicate a slightly warmer than ambient surface (approximately 1.0°C higher) and yellow to white surfaces indicate a much warmer than ambient surface (approximately 5–10°C higher).

In closing, Oxfam does believe poverty can be ended, as their research suggests that poverty is a direct function of food, energy, and income. All three of these

FIGURE 4.4   The effectiveness of insulation (Credit: bere.co.uk, February 2020.)

needs have tangible and deliverable solutions. Since a relatively small percentage of the world's population controls a relatively large percentage of the world resources (some sources claim that the richest 16% of the world's population consume 80% of the world's natural resources), this small segment of the population has tremendous control over solutions and distribution of resources. In addition to redistribution, however, reduction of inefficient use of natural resources and providing innovative solutions can result in less pressure on natural resources, reducing the impact on the planet and improving opportunities for those in need. In addition to the Oxfam Doughnut, the Human Development Index, or HDI, also takes many themes from the UN.

The HDI began in 1990 as a part of the United Nations Development Programme (UNDP). The goal of the HDI was to expand beyond the traditional economic metrics of quantifying human development through the use of gross domestic product (GDP) and instead move into measuring human well-being in a more broadly and holistically defined manner. For example, activities such as unpaid care work (the production of goods or services in a household or community that are not sold on a market), voluntary work, and creative work may not enhance people's economic perspectives, but they do increase the richness of people's lives. These efforts by the UNDP, along with the UN and Oxfam's approaches to expand the understanding of social sustainability, could be beneficial in the process of quantifying the social sustainability pillar.

The main premise behind the UNDP's work is that people and capabilities should be the ultimate criteria for assessing development, not necessarily economic growth alone (UNDP, 2015). The capabilities were originally intended to stimulate debate about government policy priorities. In short, the UNDP worked toward creating a summary of average achievement for the following dimensions of human development (measurement for the dimension in parentheses):

- Long and healthy life (life expectancy at birth)
- Being knowledgeable (mean of years of schooling for ages 25+ and expected years for children entering school)
- Having a decent standard of living (gross national income per capita)

The objective is that improvements in these three dimensions will create conditions for human development, allowing for participation in political and community lives, environmental sustainability, human security and rights, and promoting equality and social justice.

Specific groups of people are targeted in the HDI. The first group is children, as they are the future of our race and also among the most vulnerable. It is important to recognize that across the world there are diverse experiences for children, including children who are in school, children who are not in school, and children who are working (whether by choice or forced). The second group is working-age people. Again, there are multiple groups included in the discussion, including employed non-poor, unpaid care workers, working poor, unemployed, forcibly displaced, and forced labor. The third group is people older than 62 years. Like children and working-age people, older people have diverse circumstances, including people with sufficient pension, insufficient pension, or no pension. When considering social sustainability, often the differences in groups are glossed over, so classification is helpful with the development of effective social metrics.

According to the HDI, if people have an enlarged option of choices, it is possible that they could also have a higher chance of employment. By working, people potentially have significant benefits, from income and livelihood to long-term security. In addition, if women have more choices, they may be empowered in the professional workspace, which will increase participation and voice, dignity, and recognition, and most importantly for the HDI, creativity and innovation. As people are thriving in these conditions, their health could improve, the knowledge and skills could increase, and both awareness and opportunities should also increase. Once the cycle has begun, the two concepts of work and enlarged options of choices in theory will feed off each other and will both grow together. It is important to keep in mind, however, the importance of people having a solid-enough education in order to even being the positive cycle of work and enlarged option of choices. Therefore, HDI believes that increasing education must occur in order for the positive cycle to begin and thrive. In addition, HDI believes that if one of the three components is missing in a society, societal resources should be focused on that dimension.

There are several levels to sustainability and the HDI. At the most destructive end of the spectrum, the only opportunities available for people are degrading for the future and will destroy opportunities for the present. Examples of these include forced labor on deep-sea fishing boats, clearing rainforest, or trafficked workers. Moving up the spectrum, in the middle of constructive society activities, opportunities can be either limiting for the future while advancing human potential in the present or supporting opportunities for future while limiting present human potential. Examples of the first case include monocropping and/or traditional water- or fertilizer-intensive agriculture, which does help citizens today but is a mortgage of sorts for the future because of resource allocation and degradation. Examples of the second case include recycling to take pressure off natural resources but doing so without worker safeguards, protective gear, or removing contaminants, thereby threatening exposed people to health problems. Finally, at the other end of the spectrum, in ideal conditions, people could be exposed to expanding opportunities for the future while advancing human potential in the present. Examples that are being implemented today include poverty-reducing solar power or volunteer-led reforestation. These activities are not only preserving the future but also allowing for worker empowerment and increasing the standard of living in multidimensional ways not directly linked to increasing economic production.

These concepts of HDI have been refreshed in 2019 by adopting the theme: "beyond income, beyond averages, beyond today" (UNDP, 2019). The concept of inequality is heavily emphasized through all three parts of this theme. In the beyond income portion, there is a discussion of the life paths of children born in 2000. In low human development countries, children born in 2000 have a 17% chance of dying before the age of 20, an 80% chance of not having higher education, and only a 3% chance of having higher education. Those statistics can be compared to children born in 2000 in very high human development countries. In these countries, children only have a 1% chance of dying before the age of 20, a 44% chance of not having higher education, and a 55% chance of having higher education. These paths of children are very difficult to reverse and will have a significant impact on the children as they move into adulthood.

Similar examples are provided for the beyond averages and beyond today portions of the theme as well. For example, access to technology is critical for moving ahead in life. However, there are only 67.0 mobile-cellular subscriptions per 100 inhabitants in low human development group countries, whereas there are 131.6 in very high development group countries. A similar yet even more unequal trend is in fixed broadband subscriptions, where 0.8 per 100 inhabitants have fixed broadband in low human development group countries, whereas there are 28.3 in very high development group countries. This perpetuates the concept that citizens in low human development group countries will continue to diverge from citizens in high human development group countries. Lastly, the concept of gender and racial inequality is still very much in existence. Although there has been an increase of awareness through movements such as the #MeToo and #blacklivesmatter, there are still large levels of gender and racial inequality at home, in the workplace, and in politics. For example, at home, it is estimated that women perform three times more unpaid care work as men, and at the highest levels of government, there are differences of up to 90% in positions of political power. It is also estimated that black males earn 67% of that earned by white males in the United States. While these are just two statistics, there are countless others available dealing with both gender and race.

While all of these concepts within HDI have merit, implementation will need additional innovative approaches. Government policies for enhancing human development through work could include strategies for ongoing education, entrepreneurship and wealth creation, or tax policies that recognize unpaid work. This could include formulating national employment strategies aimed at preparing the national labor force to seize opportunities in the changing world of work. Also critical are strategies for ensuring workers' well-being, which include guaranteeing workers' rights and benefits, extending social protection, and addressing inequalities. To capture the benefits of these policies, however, there must be an agenda for action that implements changing the traditional mechanisms for employment protection based on principles of sustainability (a "new social contract"), mobilization of all workers, businesses, and governments around the world (a "global deal"), or formalizing employment creation and enterprise development, standards and rights at work, social protection, and governance and social dialogue (a "decent work agenda"). Strategies should be developed for targeted actions, such as reducing gender or racial inequities in the workplace, moving toward sustainable work for more adults, and undertaking group-specific initiatives to improve the well-being of marginalized populations. Finally, it is important to recognize and explore the different stages where changes can be made. Changes such as taxation are considered post-market policies, whereas inequalities while people are working are considered in-market policies, and inequalities before they start working are considered pre-market policies. Using these three policy brackets, the majority of improvements have occurred in the post-market policies, while there has been less headway in the areas of in-market and pre-market policies.

While the existing literature in civil engineering, work done by the UN and UNDP, and Oxfam have a plethora of potential metrics, there are other existing tools that have attempted to capture social impacts explicitly. Much like the popular environmental assessment tool for projects, an SIA has also been developed. This will be discussed in the following section.

Example Problem 4.3

List the potential social metrics that have been developed within the UN.

The potential social metrics that have been developed within the UN include poverty, consumption, production, human development, uneven distribution, governance, health, education, people, empowerment, removing barriers, and women.

Example Problem 4.4

List the potential social metrics that have been developed within the Oxfam Doughnut.

The potential social metrics that have been developed within the Oxfam Doughnut include education, resilience, voice, jobs, gender equality, social equality, energy, health, food, water, and income.

Example Problem 4.5

List the potential social metrics that have been developed within the HDI.

The potential social metrics that have been developed within the HDI include health, lifespan, education, and standard of living.

4.4 SOCIAL IMPACT ASSESSMENT

An SIA reviews the social effects of infrastructure projects and other development interventions. The concept was derived from the environmental impact assessment model, which traditionally evaluates environmental effects of civil engineering projects. In an SIA, there are five main types of social impacts: lifestyle, cultural, community, quality of life, and health. These social metrics expand beyond economic metrics, much like the UN and Oxfam work, and focus instead on social issue parameters. Important inputs in an SIA analysis include demographic factors, socioeconomic determinants, social organization, sociopolitical context, as well as needs and values. An SIA is performed before a project is initiated to help decide which alternative to implement.

One example of an SIA is from 2006 by the Centre for Good Governance, which developed a guide for SIA (SIA, 2006). Their definition of the term "society impacts" is "the impacts of developmental interventions on human environment." This includes ways of life, culture, community, political systems, environment, health and well-being, personal and property rights, along with fears and aspirations. As a part of this guide, multiple variables were provided to relate project stage to social impact. The four project stages are planning/policy development, implementation/construction, operation/maintenance, and decommissioning/abandonment. Each of the four project stages were associated with five SIA variables. These five variables (population characteristics, community and institution structures, political and social resources, individual and family changes, and community resources) evaluated at each project stage are intended to provide a beginning point for the SIA. Underneath these five SIA variables were specific indicators, and these are summarized in Figure 4.5.

A second example of an SIA is from Norway on a road construction project (NPRA, 2007). In this project, the socioeconomic analysis focused on nonmonetized impacts. These impacts included landscape and/or cityscape visual effects, community life, outdoor life, natural environment, cultural heritage (prehistoric deposits), and natural resources. In order to begin quantifying these impacts, values were assigned at three levels: small, medium, and large, in both the positive and negative directions. The project involved upgrading a road passing through a small town. Three different alternatives were explored: a do-nothing alternative, upgrading the existing roadway, and constructing a completely new roadway. Upgrading the new roadway suggested a positive economic effect but a negative social effect, while building a new highway suggested both a negative economic effect and a negative social effect. Noneconomic predicted benefits to upgrading the existing roadway included a neutral effect on community life, outdoor recreation, and natural resources, while the new-build roadway was predicted to have a positive effect on community life and outdoor recreation but a negative impact on natural resources. However, both options (upgrading and building new) in the SIA predicted a negative effect on landscape and cityscape, natural environment, and cultural heritage. Yet, both of these alternatives were ranked higher economically and socially versus doing nothing. Therefore, upgrading the existing roadway was ranked first and recommended for both economic and social benefits. While this example is somewhat broad, it does give some insight on the challenges to quantifying and qualifying social metrics.

When considering the existing civil engineering concepts, transportation, and social media, work done at the UN, the Oxfam Doughnut, the HDI, and SIA, it is apparent that the concept of resiliency is becoming more pronounced. In terms

Population Characteristics

- Population change
- Ethnic and racial distribution
- Relocated populations
- Influx/outflow of temporary workers
- Seasonal residents

Community and Institution Structures

- Voluntary associations
- Interest group activity
- Size and structure of local government
- Historical experience with change
- Employment/income characteristics
- Employment equity of minority groups
- Local/regional/national linkages
- Industrial/commercial diversity
- Presence of planning and zoning activity

Political and Social Resources

- Distribution of power and authority
- Identification of stakeholders
- Interested and affected publics
- Leadership capability and characteristics

Individual and Family Changes

- Perceptions of risk, health, and safety
- Displacement/relocation concerns
- Trust in political and social institutions
- Residential stability
- Density of acquaintanceship
- Attitudes toward policy/project
- Family and friendship networks
- Concerns about social well-being

Community Resources

- Change in community infrastructure
- Access to community infrastructure
- Indigenous groups
- Land use patterns
- Effects on cultural, historical, and archaeological resources

FIGURE 4.5 SIA variables and indicators

of social–ecological systems, resiliency is the capacity of the system to sustain or absorb disturbances while still being able to maintain its structure and functions. In the context of social sustainability, a resilient society would be able to overcome barriers to common tasks such as commuting, prevail over issues such as poverty and natural hazards, and develop ways to thrive by moving into the safe and just space for humanity. Individuals' self-sufficiency, or the possession of sufficient resources to survive with enough excess to be able to participate meaningfully in society, is critical to resiliency. Conquering the impediments to social sustainability will only increase the successful virtuous cycle of social enhancements along with economic enhancements. In addition to the concept of resiliency, the four broad emerging areas are highlighted in social sustainability: human well-being, access to resources, self-government, and civil society. These four emerging areas provide a framework within which earlier attempts at social sustainability metrics can be evaluated and provide a starting point for a more comprehensive set of metrics.

All of these concepts that fall under the social pillar do not often intersect with civil engineering. In order to continue the process of building social sustainability metrics for use with civil engineering, looking at the diverse portfolio of perspectives provided in this chapter can help guide the development of effective metrics. These perspectives and concepts are a beginning, and as the concepts are more widely and better understood, applications toward civil engineering processes can begin, and, in time, a suite of social metrics can be developed that are as accepted as existing economic and environmental metrics.

One perspective that is not well covered in Civil Engineering undergraduate curriculum is the business perspective. Whether designing roads or water treatment plants or analyzing a slope's stability or the deflection of a skyscraper in the wind, all engineering work is done through some sort of business. This business should care deeply about correctly analyzing engineering situations. But traditionally, this business also focused only on the economic pillar of sustainability and how the work was profitable. As discussed in Chapter 3, there has been movement toward the environmental perspective as well. Even more recently, however, there has been work done that recognizes the importance of the social purpose of corporations, or businesses, as well. This concept will be briefly introduced in the final section of this chapter.

Example Problem 4.6

List the potential social metrics that have been developed within the SIA.
    The potential social metrics that have been developed within SIA include lifestyle, cultural, community, and quality of life.

4.5   SOCIAL PURPOSE OF CORPORATIONS

While there are undoubtably dozens of different organizations working on qualifying and quantifying the social purpose of corporations, this section will focus on the British Academy. Corporations are one of the most influential institutions in today's society, with the world's 500 largest companies generating $32.7 trillion in revenue in 2018 (British Academy, 2019). Therefore, utilizing these companies as drivers of change can help address many of the challenges faced across the globe today. Therefore, the British Academy developed a case for reform of corporations to address social, political, and environmental challenges by seeking a purpose beyond maximizing profit alone. In an initial study, they found that while business leaders are increasingly using terms such as social purpose and corporate purpose and asserting the importance of having a purpose, the same leaders also expressed a need for a clear framework to articulate and measure social purpose (Hsieh et al., 2018). Therefore, the British Academy sought to develop such a framework by first distinguishing the differences between social and corporate purposes, exploring the roles social and corporate purposes fulfill in moral reflection on corporations, and finally proposing a framework for development of a purposeful business.

First, a clear definition is established for both a social purpose and a corporate purpose. Corporations are a type of legal entity which provide a framework for individuals to organize group activities. For businesses, the activities organized are directed toward producing goods and services. Often, the purpose of a corporation is believed to be limited to making profits for shareholders and owners. However, in order to profit, the good or service offered by a corporation must be valuable to at least some portion of the population. Additionally, in providing this good or service, the corporation also employs individuals, providing income, personal development, and even a sense of meaning to employees. Each of these factors contributes to the social purpose of an organization. Social purpose of a corporation is the specific contribution it makes to advancing societal goals as it relates to social systems such as politics, science, and education. Corporate purpose, on the other hand, goes beyond providing a valuable product or contributing to sustainable economic development. Therefore, corporate purpose is exclusive to corporations which are managed with the goal of achieving more than maximizing profits. Corporate purpose relates to the intentions and ambitions of a corporation rather than the measurable societal contributions. While pursuing a corporate purpose is compatible with pursuing profits, pursuit of a corporate purpose may not be the most profitable strategy and may lead to questions surrounding how various goals should be prioritized. For instance, if maximizing financial benefits conflicted with pursuing societal goals, a corporation with a strong corporate purpose may still choose to prioritize societal goals more heavily. Whereas, a corporation which does not have a clearly defined corporate purpose may choose to place societal goals on the backburner while pursuing financial gains. With this in mind, the British Academy proposed that corporate purpose become the reason a company exists and the starting point from which it operates. In this scenario, profit becomes a product of the purpose rather than the purpose itself.

Purpose of a corporation can serve three primary roles: constructive, communicative, and critical. The constructive function of purpose is invoked by either regulatory systems or the corporation itself. Regulatory systems can be developed to steer corporations toward social purpose. Corporations then develop and articulate a corporate purpose consistent with the social purpose and continually embed their corporate purpose into operations, strategy, and organizational culture. A properly formulated purpose identifies ways a corporation assists people, organizations, societies, and nations to address challenges faced while simultaneously minimizing problems a corporation may cause. Examples of statements describing corporate purposes are shown in Table 4.3. From these examples, it is apparent seeking the balance between descriptive and aspirational is important.

TABLE 4.3

Examples of Corporate Purposes (British Academy, 2019)

Corporation	Purpose
Novo Nordisk	Drive change to defeat diabetes and other serious chronic diseases
Danone North America	Expecting more from our food
Anglian Water	Conduct business and operations for the benefit of members as a whole while delivering long-term value for customers, the region, and the communities served and seeking positive outcomes for the environment and society

After establishment of a purpose, the communicative function of purpose seeks to involve multiple levels of society as corporations negotiate with governmental leaders, business regulators, and individuals within society regarding the contribution the corporation should make. This debate can begin once social and corporate purposes have been articulated and allows for all parties involved to evaluate the social harms and benefits generated by a corporation. Finally, after purposes have been developed and iterated, the critical role of purpose is holding corporations accountable for any benefit or harm created for society. In general, measures developed for purpose can be either accounting measures or action-guiding measures. For social purpose, an accounting measure may be a profit and loss statement which reviews past performance; whereas, an action-guiding measure may be a report of return on investment which compares potential projects in the planning stage to select the most beneficial alternative. Corporate purpose is typically evaluated primarily using accounting measures. An example of one measure is the Environmental, Social, and Governance (ESG) impact which will be discussed in detail in Chapter 10 (Section 4). While measures for social and corporate purpose exist, none of the existing methodologies are complete. Therefore, to fully hold corporations accountable, addressing the challenge of measuring accomplishment of purpose will be needed.

Ultimately, corporations are able to deliver on the articulated purpose through commitments to develop mutual trust with other parties impacted and instill a culture of ethics and values consistent with the purpose. Defining a purpose is the first step in a corporation becoming a trustworthy, purposeful business with an ethical culture. The next steps are outlined in eight principles for business leaders and policymakers developed by the British Academy (2019). These principles are not prescriptive; rather, they establish features of an environment which enables delivery of purpose in order to make the principles flexible to differing business models, cultures, and governments. The eight principles are outlined below and then described in greater detail in subsequent paragraphs:

1. *Corporate Law* should place purpose at the heart of the corporation and require directors to state and demonstrate commitment to purposes,

2. *Regulation* should expect particularly high duties of engagement, loyalty, and care to public interests from directors of companies which perform important public functions,
3. *Ownership* should recognize obligations of shareholders and engage them in supporting corporate purposes as well as in their rights to derive financial benefit,
4. *Corporate Governance* should align managerial interests with companies' purposes and establish accountability to a range of stakeholders through appropriate board structures, determining a set of values necessary to deliver purpose embedded in the company culture,
5. *Measurement* should recognize impacts and investment by companies in their workers, societies, and natural assets both within and outside the firm,
6. *Performance* should be measured against fulfillment of corporate purposes and profits measured net of the costs of achieving them,
7. *Corporate Financing* should be of a form and duration that allows companies to fund more engaged and long-term investment in their purposes, and
8. *Corporate Investment* should be made in partnership with private, public, and not-for-profit organizations that contribute toward the fulfillment of corporate purposes.

Principle 1 requires reformulation of corporate laws to capture both the positive benefit of producing profitable solutions to the problems of people and planet as well as avoiding harm by not profiting from producing problems for people or the planet. Corporate laws are evolving around the world to incorporate these concepts. In the United States, 36 states offer a "public benefit corporation" form, an example of which is Patagonio. In the United Kingdom, laws providing more specific obligations to avoid harm through supply chains have been developed, such as the Modern Slavery Act and the Bribery Act. A formulation which could continue to build upon these developments could require directors of companies to establish the company purpose and have regard to the consequences of any decision on the interests of stakeholders.

Regulation, as it pertains to Principle 2, is not limited to establishing rules for operating as encouraging diversity of purposes encourages innovation and ingenuity, but is meant to establish the obligation held by companies performing essential public functions to practice virtue. The significance of this principle is based in aligning the interests of company directors with the public functions the company serves. A difficulty with this principle is identifying which companies' purposes should be subject to such stipulations. An example of one company adhering to this principle is Anglian Water, a utility company in the United Kingdom. In the company's Articles of Association, the purpose statement also stipulates the director's duties include promoting the purpose of the company and carrying out operations in accordance with the statement of responsible business practices. As a result, the board of directors has a legal requirement to evaluate the impact of the company's decision on society, the environment, and their financial implications.

Principle 3 relates to ownership, challenging companies to move beyond traditional views of property rights which view ownership in light of assets to considering ownership as an obligation to respect the interests of those affected by the company's purpose. Similar to ownership of financial capital providing an incentive to shareholders to create capital, this shift in thinking would encourage the definition and support of a corporate purpose. An example is large industrial foundations in Denmark and Germany which oversee the determination and implementation of businesses' purposes and values rather than managing the companies. Employee-owned companies, such as Hershey, also illustrate a different form of ownership linked to stewardship.

Corporate governance is typically considered in the context of aligning the interests of management with those of shareholders. However, Principle 4 suggests that corporate governance should seek to align company strategy and culture with its purposes and values embedded throughout the organization. To accomplish this, German companies use a two-tier board structure which splits daily management responsibilities from strategic decision making to prioritize shareholder interests as well as employee relations.

In order for companies to succeed in pursuing each of the other principles, access to information on the performance of the corporation as it relates to purpose is essential. Therefore, measurement should provide more than an account of financial and material assets. Principle 5 states measurement needs to recognize the company's investment in workers, societies, and natural assets as well. To determine what information corporations need to measure and disclose, multiple organizations have come together under the Impact Management Project to develop standards addressing conceptual frameworks for measurement, metrics, valuation techniques, and integrated reporting. In addition to these efforts, companies should also identify firm-specific metrics which can serve to drive behavior beyond minimum regulatory standards.

According to Principle 6, corporate performance should be measured against fulfillment of purpose and profits measured net of the costs of achieving them. This idea considers business expenditure on workers, societies, and natural assets which contribute to the execution of corporate purpose to be an investment with the potential to yield benefits over time. Profit and loss statements would then recognize these expenditures as investments and expense the investments over the relevant life similar to physical assets. These new approaches, called "Integrated Profit and Loss" or "Impact Weighted Financial Accounts" seek to reflect the impact of a company on the environment, employees, and society. This impact is converted into a monetary unit allowing for profit to be adjusted for the cost of achieving or mitigating non-monetary impacts.

Along these lines, Principle 7 asserts corporate financing should allow for funding of long-term investment in a company's purpose. Incorporating this principle moves corporate financing away from only being concerned about the interests of investors. Modifying the corporate financing model also requires long-term, engaged holders of blocks of shares acting as true owners of corporate purpose. Impact Investing is one illustration of this type of investor. Impact investments are made with the intention of generating positive and measurable social and environmental impacts in addition to financial return. Federated Hermes International is an example of an investment management firm committed to responsible investing that creates long-term wealth for investors and contributes to "positive outcomes in the wider world" (Federated Hermes, 2020).

The final principle deals with corporate investment. Rather than allocating corporate investment to maximize shareholder value, investment should be made in partnership with private, public, and nonprofit organizations working toward the fulfillment of corporate purpose. This requires companies to build relationships with partner organizations and adopt the public interest in their corporate purposes. Mars Inc. seeks to build these partnerships for investment through their supply chain, employing a concept called "Economics of Mutuality." This partnership allows Mars to address issues in their supply chain around the world by mapping ecosystems, identify "pain points," identify partner organizations to help address these pain points, determine metrics to measure intervention, and establish an accounting framework to evaluate financial performance.

Each of these eight principles is interconnected and interdependent, and must be applied as appropriate for each corporation's independent scenario. In order to make the changes set out by these principles for purposeful business, five pathways forward have been identified:

- Governments must deliver legal changes,
- Businesses and investors must provide leadership,
- Feedback loops must be optimized to inform decisions and oversight,
- Partnerships must be forged to align purposes amongst and between business and stakeholder communities, and
- People must be supported to develop new skills and knowledge needed for change.

Figure 4.6 summarizes the relationships between principles and pathways. While these principles do not end the debate surrounding how purposeful business become

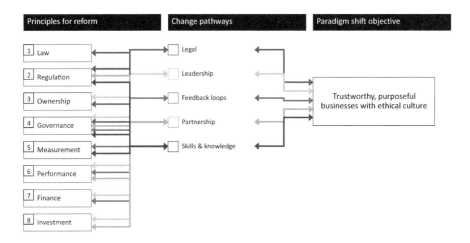

FIGURE 4.6 Relationships between principles and pathways

the norm, they provide a starting point for companies to navigate developing a clear purpose, upholding it consistently, and embedding values and a culture to support it. As more corporations adopt these principles or others which serve a similar purpose, corporations may be able to better realize their potential to solve the problems of people and planet.

Example Problem 4.7

List the eight principles for reform established by the British Academy.
   The eight principles for reform to establish a purposeful business are corporate law, regulation, ownership, corporate governance, measurement, performance, corporate financing, and corporate investment.

HOMEWORK PROBLEMS

For all Chapter 4 homework problems, use the format provided under Sidebar 1.2 "Writing a High-Quality Essay" to discuss your answer.

1. Looking at existing potential social metrics in civil engineering concepts, the UN, Oxfam, HDI, and SDI, which three metrics do you think are most important, and why?
2. Why do you think it is important to develop social metrics of sustainability? What reasons can you think for developing metrics?
3. Do you think that data can be easily found for all of the existing potential social metrics? Choose two metrics that you believe data would be easy to find, and two metrics where data would be hard to find.
4. There are societal perspectives of sustainability. World societies are sometimes divided into high income (or developed countries) and low income (developing/undeveloped countries). Focus on two existing potential social metrics and discuss how they are similar and different, based on the income level of the country.
5. Of all of the existing potential social metrics, choose a list that you think is appropriate for high-income countries and low-income countries. You will need to justify the reasons for choosing the social metrics and discuss how the data would be collected.
6. Camay et al. found that social media usage can enhance existing public participation techniques but should not entirely replace these existing techniques. Do you think over time, social media will be able to entirely replace traditional methods, such as public hearings, or will it always be recommended as a supplement?
7. Considering the example of an SIA performed for the Norway road construction project, discuss how quantifying both the economic and the social impact inform selection between design alternatives.

8. Of the eight principles for reform established by the British Academy for the realization of a purposeful business, regulation, governance, and measurement are related to the change pathway of skills and knowledge. How do these principles and pathway for change relate to ASCE's second and third priority for change (2. Standards and protocols: do the project right; 3. Expand technical capacity: transform the profession)?

REFERENCES

Alkins, A., Lane, B., Kazmierowski, T. Sustainable Pavements, Environmental, Economic, and Social Benefits of *in situ* Pavement Recycling. *Transportation Research Record: Journal of the Transportation Research Board*, No. 2084, Transportation Research Board of the National Academies, Washington, DC, 2008, pp. 100–103.

British Academy. *Principles for Purposeful Business: How to deliver the framework for the Future of the Corporation*. British Academy, Future of the Corporation, 2019.

Camay, S., Brown, L., Makoid, M. Role of Social Media in Environmental Review Process of National Environmental Policy Act. *Transportation Research Record: Journal of the Transportation Research Board*, No. 2307, Transportation Research Board of the National Academies, Washington, DC, 2012, pp. 99–107.

Chan, R., Schofer, J. Role of Social Media in Communicating Transit Disruptions. *Transportation Research Record: Journal of the Transportation Research Board*, No. 2415, Transportation Research Board of the National Academies, Washington, DC, 2014, pp. 145–151.

Cheng, J., Bertolini, L., le Clercq, F. Measuring Sustainable Accessibility. *Transportation Research Record: Journal of the Transportation Research Board*, No. 2017, Transportation Research Board of the National Academies, Washington, DC, 2007, pp. 16–25.

Douglass, J., Dissanayake, D., Coifman, B., Chen, W., Ali, F. Measuring the Effectiveness of a Transit Agency's Social Media Engagement with Travelers. *Transportation Research Record*, 2018, 2672(50), 46–55.

Evans-Cowley, J., Griffin, G. Microparticipation with Social Media for Community Engagement in Transportation Planning. *Transportation Research Record: Journal of the Transportation Research Board*, No. 2307, Transportation Research Board of the National Academies, Washington, DC, 2012, pp. 90–98.

Federated Hermes, About Us, https://www.hermes-investment.com/about-us/, 2020.

Fields, N., Cronley, C., Mattingly, S., Murphy, E., Miller, V. "You Are Really at Their Mercy": Examining the Relationship Between Transportation Disadvantage and Social Exclusion Among Older Adults Through the Use of Innovative Technology. *Transportation Research Record*, 2019, 2673(7), pp. 12–24.

Guthrie, A., Fan, Y., Das, K. Accessibility Scenario Analysis of a Hypothetical Future Transit Network: Social Equity Implications of a General Transit Feed Specification-Based Sketch Planning Tool. *Transportation Research Record: Journal of the Transportation Research Board*, No. 2671, Transportation Research Board of the National Academies, Washington, DC, 2017, pp. 1–9.

HDI. Beyond Income, Beyond Averages, Beyond Today: Inequalities in Human Development in the 21st Century. Human Development Report 2019. United Nations Development Programme, 2019.

Hsieh, N., Meyer, M., Rodin, D., Klooster, J. The Social Purpose of Corporations. *Journal of the British Academy*, 2018, 6(s1), 49–73.

Kuzio, J. Planning for Social Equity and Emerging Technologies. *Transportation Research Record*, 2019, 2673(11), 693–703.

Lucas, K., Marsden, G., Brooks, M., Kimble, M. Assessment of Capabilities for Examining Longterm Social Sustainability of Transport and Land Use Strategies. *Transportation Research Record: Journal of the Transportation Research Board*, No. 2013, Transportation Research Board of the National Academies, Washington, DC, 2007, pp. 30–37.

Mjahed, L., Mittal, A., Elfar, A., Mahmassani, H., Chen, Y. Exploing the Role of Social Media Platforms in Informing Trip Planning. *Transportation Research Record: Journal of the Transportation Research Board*, No. 2666, Transportation Research Board of the National Academies, Washington, DC, 2017, pp. 1–9.

NPRA. Impact Assessment of Road Transport Projects. Statens Vegvesen, Norwegian Public Roads Administration, Summary of Handbook 140 – Impact Assessment, 2007.

Oxfam. Fighting Inequality to Beat Poverty, Oxfam 2018–2019 Annual Report, 2019.

Pender, B., Currie, G., Delbosc, A., Shiwakoti, N. International Study of Current and Potential Social Media Applications in Unplanned Passenger Rail Disruptions. *Transportation Research Record: Journal of the Transportation Research Board*, No. 2419, Transportation Research Board of the National Academies, Washington, DC, 2014, pp. 118–127.

Raworth, K. *A Safe and Just Space for Humanity, Can We Live Within the Doughnut?* Oxfam International, Oxford, England, February, 2012.

Rockström, J. et al. Planetary Boundaries: Exploring the Safe Operating Space for Humanity. *Ecology and Society,* 2009, 14(2), pp. 472–475.

Sadri, A., Hasan, S., Ukkusuri, S., Cebrian, M. Crisis Communication Patterns in Social Media During Hurricane Sandy. *Transportation Research Record*, 2018, 2672(1), 125–137.

SIA. *A Comprehensive Guide for Social Impact Assessment.* Centre for Good Governance, Hyderabad, India, 2006.

Ukkusuri, S., Zhan, X., Sadri, A. M., Ye, Q. Use of Social Media Data to Explore Crisis Informatics: Study of 2013 Oklahoma Tornado. *Transportation Research Record: Journal of the Transportation Research Board*, No. 2459, Transportation Research Board of the National Academies, Washington, DC, 2014, pp. 110–118.

UN. *Indicators of Sustainable Development: Guidelines and Methodologies.* 3rd Edition, United Nations, New York, 2007.

UN. *Report of the United Nations Conference on Sustainable Development.* United Nations, Rio de Janeiro, Brazil, 2012.

UN. Transforming Our World: The 2030 Agenda for Sustainable Development. Resolution adopted by the General Assembly on September 25, 2015, Seventieth Session.

UNDP. *Human Development Report 2015, Work for Human Development.* United Nations Development Programme, New York, 2015.

UNDP. *Human Development Report 2019, Beyond Income, Beyond Averages, Beyond Today: Inequalities in Human Development in the 21st Century.* United Nations Development Programme, New York, 2019.

Valdes-Vasquez, R., Klotz, L. Incorporating the Social Dimension of Sustainability Into Civil Engineering Education. *Journal of Professional Issues in Engineering Education and Practice*, 2011, 137(4), 189–197.

Wojtowicz, J., Wallace, A. Use of Social Media by Transportation Agencies for Traffic Management. *Transportation Research Record: Journal of the Transportation Research Board*, No. 2551, Transportation Research Board of the National Academies, Washington, DC, 2016, pp. 82–89.

5 Application
Environmental Sustainability

It does not matter how slowly you go as long as you do not stop.

Confucius

Environmental engineering is a broad topic. The breadth of the environmental engineering topic includes chemistry, biology, ecosystems, air quality, hydraulics, hydrology, and groundwater. Environmental engineering topics also include highly engineered systems such as drinking water treatment and wastewater treatment. Other topics that may not be as intuitive, like air quality and built environment, also incorporate fundamental concepts of environmental engineering.

With such a diverse range of topics, it is not surprising that there are almost limitless applications of sustainability in environmental engineering. The environmental engineering topics discussed above are key to the UN's, 2007, 11 indicators from the Commission on Sustainable Development Sustainable Development Goals, including poverty, health, atmosphere, oceans, fresh water, and economic development. These indicators can be demonstrated through the close examination of four specific topics within environmental engineering:

1. Low impact development
2. Drinking water treatment
3. Outdoor air quality
4. Coastal resilience

These four topics will be discussed in detail in the following sections.

5.1 LOW IMPACT DEVELOPMENT

The construction of any civil engineering structure, such as a landfill, a shopping center, or an airport terminal, alters the natural landscape. Existing vegetation is generally removed, the natural topography is modified, and new structures are installed. These types of activities are called development, and this development modifies the hydrological cycle at a local level. Instead of water falling onto the ground surface and naturally absorbing into the soil, it is often collected and diverted to retention ponds or storm sewers. These types of large water diversion systems not only move significant volumes of water to alternative locations, but they are also expensive to construct and maintain. A potential alternate to constructing these systems is to explore alternate methods of design that have a lower impact on the ecosystems. These methods are called low impact development, or

LID. As discussed above, one of the primary considerations of LID in civil engineering is runoff. The diversion of runoff has two downsides in addition to cost. First, this water often picks up pollutants and carries them downstream, creating smaller areas of higher concentrations of pollutants, and overloads nature's ability to filter pollutants from water. Second, the diversion of water can reduce the groundwater table, as water that would originally have percolated downward and naturally recharged the water table instead is transferred to another location. At some point during rain events, there will be a maximum amount of water moving across a field, a parking lot, or any other natural or engineered surface. This maximum amount of water is called the peak runoff.

There are two common methods of quantifying the peak runoff, which can be used for sizing culverts and storm drains. First, the rational method (also known as the rational formula, the rational equation, or the Lloyd-Davies equation) is typically used for relatively small areas, generally less than one-half square mile. The peak discharge is calculated from the rational method from Equation 5.1:

$$Q = C \times I \times A \tag{5.1}$$

where
$\quad$ Q $\;=\;$ peak discharge (ft³/sec)
$\quad$ C $\;=\;$ runoff coefficient
$\quad$ I $\quad=\;$ rainfall intensity (in/hr)
$\quad$ A $\;=\;$ watershed area (acres)

There are two important points to highlight in Equation 5.1. First, units must be carefully followed, as the runoff coefficient includes conversion factors from acres and inches to cubic feet, and hours to seconds. Second, there are many sources available for runoff coefficients, and usually there is a range associated with each surface and use as well. Table 5.1 shows various examples of typical values of runoff coefficients for the rational method compiled from various sources.

An important takeaway from Table 5.1 is the variability of runoff coefficients. Not only do many assumptions need to be made when choosing a runoff coefficient, but the calculation of runoff coefficient is not standard as well. As engineers, it is important to state the assumptions and choose a conservative yet reasonable value of runoff coefficients when performing a design.

A second method of calculating runoff, which can be applied to any size homogeneous watershed, is the NRCS rainfall-runoff method. The US Natural Resources Conservation Service (NRCS) calculation method has been correlated to actual experience, and revolves around the concept of a curve number, which characterizes the land use and soil type. There are several assumptions necessary before applying the NRCS method. First, the method assumes that the initial abstraction (depression storage, evaporation, and interception losses) is equal to 20% of the maximum basin retention. Second, the precipitation must equal or exceed the initial abstraction. Third, the storage capacity must be large enough to absorb the initial abstraction plus any infiltration. Fourth, the method assumes a type II storm, which is the most

TABLE 5.1

Runoff Coefficients for the Rational Method

Surface	PE Reference Manual[1]	City of Fayetteville[2]	Washington State[3]	Florida State[4]
Forest	0.059–0.20	0.15–0.30	0.10–0.30	0.10–0.30
Lawn–sandy soil <2% slope	0.05–0.10	0.15	0.05–0.10	0.10–0.15
Lawn–sandy soil 2–7% slope	0.10–0.15	0.25	0.07–0.15	0.20–0.25
Lawn–sandy soil >7% slope	0.15–0.20	0.30	0.10–0.20	0.25–0.35
Lawn–clay soil <2% slope	0.13–0.17	0.35	0.10–0.17	0.20–0.25
Lawn–clay soil 2–7% slope	0.18–0.22	0.40	0.15–0.22	0.25–0.30
Lawn–clay soil >7% slope	0.25–0.35	0.45	0.20–0.35	0.30–0.40
Asphalt	0.70–0.95	0.95	0.90	0.95
Brick	0.70–0.85	0.85	0.90	0.75–0.95
Concrete	0.80–0.95	0.95	0.90	0.95
Shingle roof	0.75–0.95	0.95	0.90	0.95
Driveways, walkways	0.75–0.85	0.95	0.75–0.85	0.95

[1] – from Lindeburg, M. Civil Engineering Reference Manual, 9th Edition, Professional Publications, Inc., 2003.

[2] – from City of Fayetteville. Drainage Criteria Manual, Fayetteville, Arkansas, 2014.

[3] – from Washington State. Hydraulics Manual, Washington State Department of Transportation, 1997.

[4] – from Florida State. Drainage Handbook Hydrology, State of Florida Department of Transportation, 2012.

common type of rain event in the United States. Finally, fifth, the soil condition is assumed to be average (ARC II). The runoff is calculated using the following:

$$Q = \frac{(P - I_a)^2}{(P - I_a + S)} = \frac{(P - 0.2S)^2}{(P + 0.8S)} \tag{5.2}$$

$$S = \frac{1,000}{CN} - 10 \tag{5.3}$$

where

Q = runoff (in)
I_a = initial abstraction (in)
P = precipitation (in)
S = maximum basin retention (in)
CN = curve number (ft³/sec)

Similar to the rational method, units must be carefully followed in Equations 5.2 and 5.3 to ensure that the final runoff answer is in inches. Table 5.2 shows some typical curve numbers for the same surfaces outlined in Table 5.1

TABLE 5.2
Runoff Curve Numbers (CN, ft³/sec) of Urban Areas from NRCS

Surface	Soil group A	Soil group B	Soil group C	Soil group D
Forest	30–45	55–66	70–77	77–83
Lawn	39–68	61–79	74–86	80–89
Asphalt	98	98	98	98
Brick	98	98	98	98
Concrete	98	98	98	98
Shingle roof	98	98	98	98
Driveways, walkways	98	98	98	98

TABLE 5.3
Hydrological Soil Group Classification

Group	Infiltration rate (in/hr)	Urbanized Classification
A	>0.30	Sand, loamy sand, sandy loam
B	0.15–0.30	Silty loam, loam
C	0.05–0.15	Sandy clay loam
D	<0.05	Clay loam, silty clay loam, sandy clay, silty clay, clay

The hydrological soil group is determined by infiltration rate, with values and potential applications shown in Table 5.3.

Using these two methods of quantifying the peak runoff, several potential sustainable applications can be implemented in order to reduce peak runoff, so as to direct more of the rainfall directly downward in order to recharge the water table, reducing the need for built structures to redirect water. The applications discussed below include green roofs, porous pavements, and bioretention cells.

Example Problem 5.1

A forest in the City of Fayetteville will be replaced by a brick parking lot. Using runoff coefficients from the City of Fayetteville, calculate the change of peak discharge if the design rainfall intensity is 2.0 in/hr and the area of the brick parking lot is designed to be 10,000 ft². State any assumptions that are needed to complete this calculation.

Equation 5.1, the rational method, is used to solve the peak discharge. Using Table 5.1, it is assumed that the runoff coefficient falls in the middle of the runoff coefficient values, so 0.225 is used. The runoff coefficient for the bricks is given in Table 5.1 as 0.85. The rainfall intensity is given at 2.0 in/hr (the correct units), but the area give must be converted to the proper units of acres:

$$\frac{10{,}000 \text{ ft}^2}{1} \times \frac{2.2957e^{-5} \text{ acres}}{1.0 \text{ ft}^2} = 0.2296 \text{ acres}$$

Next, Equation 5.1 can be used to calculate the peak discharge for each surface:

$$\text{Forest} \rightarrow Q = C \times I \times A = 0.225 \times 2.0 \frac{\text{in}}{\text{hr}} \times 0.2296 \text{ acres} = 0.103 \text{ ft}^3/\text{sec}$$

$$\text{Brick parking lot} \rightarrow Q = C \times I \times A = 0.85 \times 2.0 \frac{\text{in}}{\text{hr}} \times 0.2296 \text{ acres} = 0.390 \text{ ft}^3/\text{sec}$$

$$\text{Difference} \rightarrow 0.390 \text{ ft}^3/\text{sec} - 0.103 \text{ ft}^3/\text{sec} = 0.287 \text{ ft}^3/\text{sec} \text{ increase in discharge}$$

5.1.1 GREEN ROOFS

The tops of buildings are designed to be impermeable, so water does not infiltrate into the built structure. However, this impervious surface directs water away from where it would naturally absorb into the earth, decreasing the water table beneath the building and increasing water flow away from the building. While a single structure is not likely to significantly influence an area's water table or runoff, the more urbanized an area becomes, the more significant the problems associated with the impervious surfaces are. Green roofs are a potential solution to these problems. Green roofs consist of multiple layers of natural vegetation, synthetic material, and impermeable material. The natural vegetation sits on the surface of the structure, with a filter and drainage layer directly underneath. With proper design, the drainage from the roof can be managed so that the majority of water is directed downward into the water table directly underneath the building, as opposed to flowing to another area. Finally, between the drainage layer and the structure's roof is a protection layer and root barrier to protect the structure itself. Figure 5.1 shows a green roof on top of

(a)   (b)

FIGURE 5.1 a) University of Arkansas Hillside Auditorium green roof (credit: A Braham). b) University of Arkansas Hillside Auditorium, green roof foreground, traditional roof background on the Mechanical Engineering building (credit: A Braham)

Hillside Auditorium at the University of Arkansas. The green roof is over one of the two auditoriums inside the building. It is interesting to note in Figure 5.1a that a standard roof can be seen in the background (the Mechanical Engineering building at the University of Arkansas), complete with artificially engineered drainage to handle runoff.

The General Services Administration, or GSA, has been a leader of green roof implementation in the United States (GSA, 2011). The GSA has categorized green roofs into four main groups: single course extensive, multi-course extensive, semi-intensive, and intensive. These categories are dependent on the thickness, the type of drainage layer, the type of vegetation layer, and the media type. The choice of green roof is dependent on the local environment, the level of management that the owner is willing to engage in, and the structural capacity of the structure that the green roof will sit on. Table 5.4 summarizes the four types of green roofs, and Figure 5.2 shows typical cross sections of each type.

With the relatively small footprint of each green roof, the rational method is often employed in order to calculate the difference in runoff for an area. Therefore, runoff coefficients need to be determined for green roofs. Several agencies and research groups have investigated potential runoff coefficients for green roof systems. Table 5.5 summarizes samples of these values. Two trends are apparent from Table 5.5. First, the runoff coefficients of green roofs are all lower than the impervious roofs

TABLE 5.4
Summary of Types of Green Roof Systems

	Single-course extensive	Multi-course extensive	Semi-intensive	Intensive
Thickness (in)	3–4	4–6	6–12	>12
Drainage layer	Moisture management layer	Based on growth media thickness, plants, and local climate	Discrete drainage layer	Discrete drainage layer
Vegetation Layer	Sedum, other succulents	Sedum, other succulents	Meadow species, ornamental varieties, woody perennials, turf grass	Similar to ground level
Media type	Coarse	Finer grained	Multi-course	Intensive growth media
Irrigation	None	First year only	Required for turf grass	Required
Prevalence	Common internationally	Most common in the United States	Common internationally	Least common, structural capacity limiting

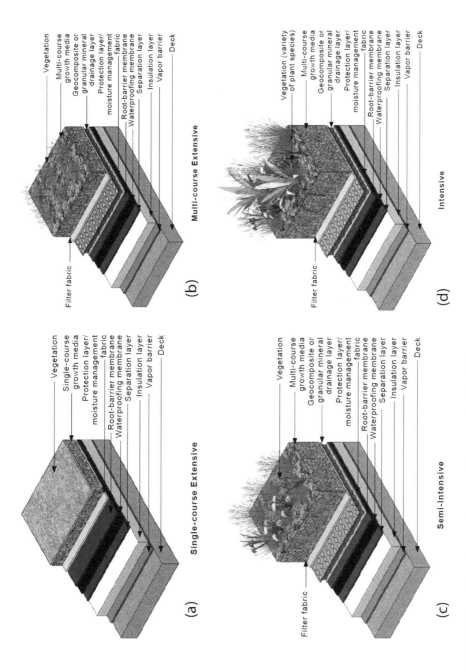

FIGURE 5.2 Examples of types of green roof systems (credit: US General Services Administration)

TABLE 5.5
Green Roof System Runoff Coefficients for the Rational Method

Source	Description	Runoff coefficient
NYC, 2012	Green roof with four or more inches of growing media	0.70
Morian *et al.*, 2005	Average soil media depth three inches	0.50
Mobilia et al., 2014	Layer depth of 15 cm, roof slope lower than 15 degrees	0.35

(Table 5.1, 0.75–0.95). Second, there is a significant range depending on the study and the roof structure and roof geometry. However, this uncertainty is not uncommon in engineering and must be considered in design.

Example Problem 5.2

The University of Arkansas is planning on replacing the traditional shingle roof with a green roof that has an average soil media depth of 3 inches. The approximate roof area of the Mechanical Engineering building is 2,576 yd². Assuming the rainfall intensity is 1.8 in/hr, what is the change in peak discharge? Use the PE Reference Manual runoff coefficient for the shingle roof. State any assumptions that you made.

Equation 5.1, the rational method, is used to solve for the peak discharge. Using Table 5.1, it is assumed that the runoff coefficient for the shingle roof falls in the middle of the runoff coefficient values, so 0.85 is used. The runoff coefficient for the green roof with an average soil media depth of 3 inches is given in Table 5.5 as 0.50. The rainfall intensity is given at 1.8 in/hr (the correct units), but the area give must be converted to the proper units of acres:

$$\frac{2,576 \text{ yd}^2}{1} \times \frac{2.066e^{-4} \text{ acres}}{1.0 \text{ yd}^2} = 0.5322 \text{ acres}$$

Next, Equation 5.1 can be used to calculate the peak discharge for each surface:

$$\text{Shingle roof} \rightarrow Q = C \times I \times A = 0.85 \times 1.8 \frac{\text{in}}{\text{hr}} \times 0.5322 \text{ acres} = 0.814 \text{ ft}^3/\text{sec}$$

$$\text{Green roof} \rightarrow Q = C \times I \times A = 0.50 \times 1.8 \frac{\text{in}}{\text{hr}} \times 0.5322 \text{ acres} = 0.479 \text{ ft}^3/\text{sec}$$

$$\text{Difference} \rightarrow 0.814 \text{ ft}^3/\text{sec} - 0.479 \text{ ft}^3/\text{sec} = 0.335 \text{ ft}^3/\text{sec decrease in discharge}$$

5.1.2 Porous Pavements

Another sustainable application to reduce peak runoff is the use of porous, also called permeable, pavements. According to the Federal Highway Administration, there are 8.5 million lane-miles (13.7 million lane-kilometers) of road in the United States (FHWA, 2011). Assuming that each lane is 12 ft wide (3.65 meters), approximately 19,300 square miles (49,320 square kilometers) of roadway cover the United States. This is equivalent to approximately 9.35 million American football fields including the end zones. While soccer pitches are not a standard size, using the preferred size of 105 × 68 meters, this is equal to 6.91 million soccer pitches. This immense amount of space provides an excellent opportunity to decrease the amount of impervious surfaces by using porous pavements.

Porous pavements typically come in three forms: porous asphalt, porous concrete, or pavers. Porous asphalt and porous concrete utilize gap-graded aggregate gradations with air voids typically between 15% and 25%. This allows for more water to pass through the pavement layer instead of running off the pavement layer. Additional benefits include filtering of the water and the potential to reduce heat island effect (the increase of temperature in urban areas due to the dark color of engineered structures). With the unique gradation and high air voids, porous pavements are not intended for high-volume or high-load roadway applications, so use on interstates or large state highways is not advised. However, the pavement surface is more than appropriate for lower volume roads and residential areas, which make up the vast majority of roadways in the United States.

In brief, porous pavements are typically built over uncompacted subgrades. This allows for natural infiltration into the existing groundwater table. Porous pavements can be placed in a typical pavement structure, but can also be placed on a choker course, stone reservoir, and a geotextile fabric. These reservoirs allow for the storage of water during storm events, allowing for the natural seepage of water down into the groundwater. Typically, the reservoirs are uniformly graded, clean crushed stone, with up to 40% voids. Immediately above the reservoir is the choker course, which provides a stabilized surface for the bound layer above. The geotextile is placed between the natural soil and reservoir in order to prevent the migration of fines up into the pavement layers. Figure 5.3 shows an example of a porous asphalt pavement structure.

There are three key considerations when designing the thickness of the porous pavement layers. The first consideration is the site of the project, which includes depth of bedrock, soil types, and pavement slope. The second consideration is the hydrology design, as the amount of storage needs to be adequate for anticipated precipitation and edge drains may be required to prevent the surface layers from overflowing. Finally, the structural design must be addressed. Utilizing the AASHTO 1993 Design Guide for flexible pavements, typical asphalt concrete structural number coefficients generally are around 0.44. However, porous asphalt has slightly lower structural number coefficients, generally in the range of 0.40–0.42 (Hansen, 2008). This decrease in capacity needs to be taken into account during structural design.

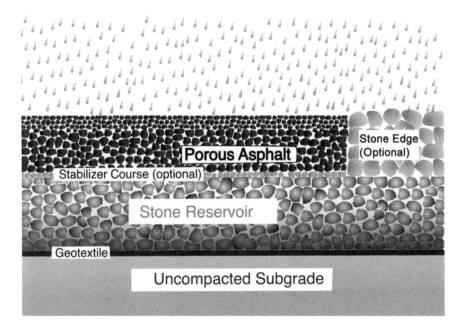

FIGURE 5.3   Potential porous asphalt pavement structure with a stone reservoir and geotextile layer (credit: National Asphalt Pavement Association)

Several agencies and research groups have investigated potential runoff coefficients and curve numbers for porous pavement systems. Table 5.6 summarizes a sample of runoff coefficient values, and Table 5.7 summarizes a sample of curve numbers. Two trends are apparent from Table 5.6. First, the runoff coefficients of porous pavements are all lower than traditional roadways (Table 5.1, asphalt street = 0.70–0.95; concrete street = 0.8–0.95). Second, there is a significant range depending on the study and the roadway structure and roadway geometry. Again, this uncertainty is not uncommon in engineering and must be considered in design.

TABLE 5.6
Porous Pavement Runoff Coefficients for the Rational Method

Source	Description	Runoff coefficient
Fassman and Blackbourn, 2010	10–90th percentile events	0.29–0.67
Fassman and Blackbourn, 2010	Porous pavement about one-half otherwise impervious catchment	0.41–0.74
St. John, 1997	Newly installed porous pavement	0.12–0.40
Wei, 1986	3–4 years after installation, porous pavement	0.18–0.29

TABLE 5.7
Porous Pavement Curve Numbers for Events Greater than 50 mm Using the NRCS Method (Bean *et al.*, 2007)

Pavement system	Description	Curve number (range, mean)
Concrete grid pavers (CGP), 90 mm thick, filled with coarse grain sand	Slope 0.5%; above 50mm bedding sand, geotextile, 70mm washed marlstone	41–98, 70
Porous concrete (PC), 200 mm thick	Slope 0.3%; directly on native fine graded sand	60–91, 77
Permeable interlocking concrete pavements (PICP), 75 mm UNI Eco-Stone Pavers	Slope 0.4%; unlined; 75 mm No. 72 pea gravel, 200 mm No. 57 washed gravel	37–50, 43

Example Problem 5.3

The University of Arkansas is considering replacing Parking Lot 71, which is currently a traditional asphalt surface, with a 200-mm thick porous concrete pavement. Assuming that the old pavement will be completely removed and that the porous concrete pavement will be placed on native fine graded sand, and the precipitation is 2.6 inches, what is the change in runoff? State any assumptions that are needed to complete this calculation.

Equation 5.2, the NRCS rainfall-runoff method, is used to solve for the runoff. Using Table 5.2 for the asphalt surface (CN = 98 ft³/sec) and assuming the average from Table 5.7 for the porous concrete (CN = 77 ft³/sec), the runoff can be calculated. First, the maximum basin retention is calculated for each surface using Equation 5.3:

$$\text{Asphalt surface} \rightarrow S = \frac{1{,}000}{CN} - 10 = \frac{1{,}000}{98 \ \text{ft}^3/\text{sec}} - 10 = 0.204 \ \text{in}$$

$$\text{Porous concrete surface} \rightarrow S = \frac{1{,}000}{CN} - 10 = \frac{1{,}000}{77 \ \text{ft}^3/\text{sec}} - 10 = 2.987 \ \text{in}$$

Next, the runoff can be calculated using:

$$\text{Asphalt surface} \rightarrow Q = \frac{(P-0.2S)^2}{(P+0.8S)} = \frac{(2.6-0.2\times0.204)^2}{(2.6+0.8\times0.204)} = 2.37 \ \text{in}$$

$$\text{Porous concrete surface} \rightarrow Q = \frac{(P-0.2S)^2}{(P+0.8S)} = \frac{(2.6-0.2\times2.987)^2}{(2.6+0.8\times2.987)} = 0.80 \ \text{in}$$

$$\text{Change in runoff} \rightarrow 2.37 \ \text{in} - 0.80 \ \text{in} = 1.57 \ \text{in decrease in runoff}$$

5.1.3 BIORETENTION CELLS

The third application discussed to decrease peak runoff, directing more rainfall directly downward to recharge the water table, and reducing the need for built structures to redirect water, is the concept of bioretention cells. Bioretention cells are areas of plants and other porous materials placed near impervious surfaces that allow for the collection, filtration, infiltration, and recharge of water that runs off an impervious surface. For example, when a parking lot is constructed, a large area of land is generally covered by a relatively impervious surface. The runoff from this surface can be directed to a bioretention cell. Once at the cell, the runoff can be filtered and stored so there is the slow recharge of the groundwater table. Figures 5.4a–c show several views of a parking lot at the University of Arkansas specifically designed to accommodate bioretention cells. These views can be contrasted to Figure 5.4d,

(a)   (b)

(c)   (d)

FIGURE 5.4 a) Water can run directly from parking spaces into a bioretention cell since curbs are not installed. b) In some locations, curbs direct water to openings that flow to the bioretention cell. Note the sidewalk required a small bridge so pedestrians can cross water flow during heavy rains. c) Curbs can divert water to the bioretention cell. d) A traditional drain leading to the storm sewer, where water is diverted from one location and moved to an alternate location.

which shows a "standard" drainage solution, involving high volumes of runoff and engineered systems to divert excess water to the storm sewer.

In general, there are four different types of bioretention cells: infiltration/recharge, filtration/partial recharge, infiltration/filtration/recharge, and filtration only. Infiltration/recharge cells are most useful in areas that require a high level of recharge of water, but the in-place soil must be able to accommodate the inflow levels. The term recharge is used when water moves directly downward from the ground surface into the groundwater. A filtration/partial recharge cell is used where there is a need for a high level of filtration of the water. Therefore, the plant type is critical to match the type of pollutants expected, and a drain is used in order to aid with controlling runoff, as all of the water is not designed to recharge the groundwater table. The third type of cell is an infiltration/filtration/recharge cell. This type of cell is used when high nutrient loadings, for example, nitrates, may be present along with the ability of the water to recharge the groundwater table. In this configuration, a raised drain is available in case of several water loadings, but extra reservoir space is designed for the ability to store water for maximum recharging. Finally, the forth type of cell is utilized for areas that are known to have a high chance of significant runoff, such as gas stations, transload facilities, and transportation depots. Here, an impervious liner is used at the bottom of the system to ensure that a minimum amount of groundwater contamination would occur in a spill situation.

While bioretention cells are a viable component of Low Impact Development, they do not have specific runoff coefficients associated with them. An interesting concept, however, is the blend of different development types, or areas, and how bioretention cells can be incorporated into such spaces. In general, the size of the bioretention area is a function of the drainage area and runoff from the area. If the bioretention cell has a sand bed, the area of the cell should be approximately 5% of the drainage area multiplied by the runoff coefficient (EPA, 1999). If the bioretention cell does not have a sand bed, the area of the cell should be approximately 7% of the drainage area multiplied by the runoff coefficient.

Example Problem 5.4

The University of Washington athletic department is considering placing several bioretention cells in the asphalt parking lot directly adjacent to the football stadium (Husky Stadium). Assume the parking lot has 20,000 ft^2 of pavement and use the Washington State runoff coefficient. If the athletic department would like to use a sand bed bioretention cell, what is the required square footage of a sand bed bioretention cell?

First, use Table 5.1 to determine the runoff coefficient for asphalt, which is 0.90. Next, multiply the runoff coefficient by 20,000 ft^2 to obtain 18,000 ft^2. Finally, the sand bioretention cell should be designed as 5% of the drainage area multiplied by the runoff coefficient. Therefore, 5% of 18,000 ft^2 is (18,000 × 0.05) = 900 ft^2 of sand bed bioretention cell required for the parking lot.

5.2   DRINKING WATER TREATMENT

According to the World Health Organization (WHO, 2013), humans require 2.5–3.0 liters/day (0.7–0.8 gallons/day) for survival, 2–6 liters/day (0.5–1.6 gallons/day) for basic hygiene practices, and 3–6 liters/day (0.8–1.6 gallons/day) for basic cooking needs. For the water that is consumed uncooked, the treatment is required, as it may contain harmful microorganisms and organic or inorganic compounds that can cause physiological effects or negatively affect the taste. Evaluating drinking water can be split into two general categories: physical characteristics and contaminant regulation.

The physical characteristics of natural water include turbidity, particles, color, taste and odor, and temperature. Turbidity refers to the optical clarity of the water, and is reported in nephelometric turbidity units (NTU). Particles are solids (often not seen by the naked eye) that are suspended (> 1 μm), colloidal (0.001–1 μm), and dissolved (< 0.001 μm). Color is often influenced by dissolved organic matter, metallic ions, and turbidity, while taste and odor are generally from dissolved natural organic or inorganic constituents. In 1974, the UN Congress passed the Safe Drinking Water Act, which placed the responsibility of setting water quality regulations onto the Environmental Protection Agency, or EPA. After treatment, the water is potable and is safe for human consumption.

The regulated drinking water contaminants include microorganisms, disinfectants, disinfection by-products, inorganic chemicals, organic chemicals, and radionuclides. Each of these contaminants has a maximum contaminant level goal (MCLG) and maximum contaminant level (MCL). The MCLG is set so there is no known or expected risk to health, while the MCL is the highest level of contaminant that is allowed in drinking water. In general, the MCLs are set as close to the MCLGs as possible, balancing the best technology with cost. Table 5.8 summarizes a sampling of contaminants and their MCLG and MCL according to the US Environmental Protection Agency (EPA, 2009).

There are many different ways to quantify the sustainability of water treatment technologies. The activated carbon adsorption, air stripping, clarifier/sedimentation basin design, settling characteristics of contaminants, softening mechanisms, flocculation design, osmosis, ultrafiltration, or disinfection all have been analyzed for sustainable technologies. As an example, the Langmuir isotherm, as a part of activated carbon adsorption, will be examined in more detail here.

The adsorption process can take place in either fixed-bed filtration units or suspended-media contactors. In the fixed-bed geometry, the water passes through 1–3 meters of media (in this example, activated carbon). In the suspended-media contactors, the media is mixed with the water and travels with the water through the treatment plant. The media is usually removed by either sedimentation or filtration. During the adsorption process, various contaminants usually associated with taste and odor are transferred from the water to the media. When activated carbon is the media, the average diameter of activated carbon particles is 0.5–3.0 mm (granular particles) for fixed-bed, while the average diameter is 20–50 μm (powdered particles) for suspended media.

TABLE 5.8
Sampling of Drinking Water Standards from the EPA

Contaminant	Maximum contaminant level (MCL)	Maximum contaminant level goal (MCLG)
Asbestos (inorganic chemical, fibers >10 micrometers)	7 million fibers/liter	7 million fibers/liter
Benzene (organic chemical)	0.005 mg/L	0.0 mg/L
Chlorine as Cl$_2$ (disinfectant)	Maximum Residual Disinfectant Level = 4.0 mg/L	Maximum Residual Disinfectant Level Goal = 4.0 mg/L
Chlorite (disinfection by-product)	1.0 mg/L	0.8 mg/L
Lead (inorganic chemical)	Less than 10% of tap water samples contain less than 0.015 mg/L	0.0 mg/L
Uranium (radionuclides)	30 µg/L	0.0 µg/L
Total coliforms, positive samples per month (microorganism)	5.0%	0.0 %

Adsorption is an equilibrium process. In order to achieve and maintain equilibrium, the adsorbate is distributed between the aqueous and solid phases according to the adsorption isotherm. This is achieved by balancing the adsorbent surface (activated carbon) fixed sites where molecules of adsorbate (the contaminant) may be chemically bound. The Langmuir isotherm explains the variation of adsorption with pressure. In order to execute this isotherm, five assumptions need to be met:

- A fixed number of vacant (or adsorption) sites are available on the surface of the solid,
- All of the vacant sites are the same size and shape,
- Each site can hold one gaseous molecule and a constant amount of heat energy is released during the process,
- There is a dynamic equilibrium between adsorbed gaseous molecules and the free gaseous molecule, and
- Absorption is unilayer or monolayer.

The Langmuir isotherm can be represented using Equation 5.4:

$$\frac{x}{m} = X = \frac{aKC_e}{1 + KC_e} \tag{5.4}$$

where:
x = mass of solute adsorbed
m = mass of adsorbent
X = mass ratio of the solid phase/mass of adsorbed solute per mass of adsorbent

a = mass of adsorbed solute required to saturate completely a unit mass of adsorbent

K = experimental constant

C_e = equilibrium concentration of solute, mass/volume

It is often quite convenient to portray the Langmuir isotherm in terms of the maximum sorption capacity, which is a function of both K and a. If this is done, Equation 5.4 can be rearranged to Equation 5.5:

$$\frac{1}{q} = \frac{1}{Q_{max}} + \frac{1}{(Q_{max}) \times b \times C_e} \tag{5.5}$$

where

q   = amount of metal sorbed at equilibrium (mg/g)

Q_{max} = maximum sorption capacity of system (mg/g)

b   = constant related to binding energy of sorption system (1/mg)

C_e = concentration of metal solution at equilibrium (mg/l)

The process required to produce activated carbon from virgin raw carbon requires several energy-intensive stages, including heating the carbon up to temperatures greater than 500°C several times during the production process. Therefore, it is advantageous to explore alternative materials to replace virgin carbon in hopes of reducing raw material usage and production energy. Various examples of agricultural byproducts will be reviewed that discuss using the Langmuir isotherm in sustainable applications.

The first example examined the use of rice hulls for the sorption of cadmium (Kumar and Bandyopadhyay, 2006). Table 5.9 shows various examples of sorption capacities of not only four rice hulls, but also other agricultural by-products, and the more traditional granular or powdered activated carbon. The four types are:

- RRH – raw rice husk
- NRH – NaOH-treated rice husk
- ARH – acid-treated rice husk
- NCRH – sodium carbonate–treated rice husk

Modifying RRH improved the sorption capacity by 3–12 mg/g, and while NCRH and NRH both increased the uptake capacity, NCHR was potentially the best because of the relative low cost of sodium bicarbonate (three times less than NaOH) and the rapid uptake, where short contact time is common.

The second example utilized sunflower stalks as the raw material (Sun, 1998). The stalks were evaluated with three metals, two size ranges, and two temperatures. Table 5.10 shows the results.

These two studies demonstrate that the adsorption capacity of contaminents is influenced heavily by not only the type of media (whether coal based or agricultural waste product based), but also the contaminent being removed, the size of the media, and the temperature of the filtration.

TABLE 5.9
Maximum Sorption Capacity of Cd(II) of Various Agricultural By-products and Activated Carbon (Cd(II) EPA MCL = 0.005 mg/L)

Material	Sorption capacity (Q_{max}, mg/g)	Material	Sorption capacity (Q_{max}, mg/g)
Peanut hulls	5.96	Corncobs	8.89
Bark	8.00	Cornstarch	8.88
Powdered activated carbon	3.37	Granular activated carbon	3.37
Sawdust	9.26	Sugar beet pulp	17.2
Spent grain	17.3	RRH	8.58
Exhausted coffee	1.48	ERH	11.12
Sheath of palm	10.8	NCRH	16.18
		NRH	20.24

TABLE 5.10
Maximum Sorption Capacity of Sunflower Stalks

Metal ions contaminent	Size (mesh, openings/inch)	Temperature (°C)	Sorption capacity (Q_{max}, mg/g)
Cu^{2+}	25–45	25	25.11
(EPA secondary* =		50	24.75
1.0 mg/L)	< 60	25	29.30
		50	27.57
Zn^{2+}	25–45	25	27.27
(EPA secondary =		50	10.06
5.0 mg/L)	< 60	25	30.73
		50	11.61
Cd^{2+}	25–45	25	34.85
(EPA MCL =		50	27.24
0.005 mg/L)	< 60	25	42.18
		50	30.86
Cr^{3+}	25–45	25	15.20
(EPA MCL =		50	25.07
0.10 mg/L)	< 60	25	15.16
		50	21.48

* EPA secondary indicates a non-enforceable guideline regarding contaminants that may cause cosmetic effects (skin/tooth discoloration) or aesthetic effects (taste, odor, color), these limits are considered "secondary maximum contaminant levels."

Example Problem 5.5

Washington County, in Northwest Arkansas, is evaluating alternative materials from powdered activated carbon (with an assumed value of b = 0.237) in order to adsorb contaminants from drinking water. Since there is a significant amount of rice grown in southern Arkansas, they would like to explore using raw rice husk (with an assumed b value of 0.0496) as an alternative to powdered activated carbon. If the water currently has a level of 0.21 mg/L of Cd(II), how much powdered activated carbon and raw rice husk must be used in order to treat one liter of water to meet the EPA maximum?

First, let's explore the powdered activated carbon. To calculate q, the amount of metal sorbed at equilibrium, Equation 5.5 is used:

$$\frac{1}{q} = \frac{1}{Q_{max}} + \frac{1}{(Q_{max}) \times b \times C_e} = \frac{1}{3.37} + \frac{1}{3.37 \times 0.237 \times 0.21} = \frac{1}{6.253}, \text{ so } q = 0.16 \frac{mg}{g-L}$$

Next, dividing the difference between the actual amount of Cd(II) and the desired amount (0.005 mg/L, given in Table 5.9) by q will give the amount of activated carbon needed per liter of water:

Amount of activated carbon needed = (0.21–0.005)/(0.160) = 51.3 g of mateiral/L

Second, we use the same procedure to calculate the amount of raw rice husk needed. Using the unique values of Qmax and b, the amount of raw rick husk needed is 96.4 g of material/L.

These values are high and demonstrate why activated carbon or alternate materials are rarely used to remove metals. Metals are typically removed by ion exchange or the precipitation process, where these types of materials are generally used to remove organics (such as pesticides and VOCs). Regardless, it is useful to see an example of how alternate materials can be utilized.

5.3   OUTDOOR AIR QUALITY

The first instance of pollution probably occurred when households began lighting fires indoors for heating the air and cooking food. Indoor fires with no release vents create highly unpleasant conditions, thus chimneys were created to divert the pollution out of the house and into outdoor air. In addition, very small point source pollution can occur from something as seemingly inconsequential as a personal grill in the backyard, as seen in Figure 5.5.

While the atmosphere has an incredible high capacity to disperse pollutants, the arrival of the industrial revolution, and creation of factories, provided a pivot point in the history of air quality. With the combination of residential, commercial, and industrial pollution sources, the pollution began to outpace the ability of the atmosphere to disperse the pollutants. The problem was compounded by in exponential growth of energy consumption. During the Industrial Revolution, most power was provided by coal power plants. Thus, the pollution from coal power plants for electricity and the pollution from residential, commercial, and industrial sources caused significant air quality issues. For example, the most famous instance of London

FIGURE 5.5   Emissions from grilling hamburgers in the backyard

Fog occurred in 1879, where for four months, the sun could not be seen because of the heavy pollution. Fogs continued into the 1950s, where in 1952, approximately 4,000 London citizens were killed in a four-day fog. This event caused the English Parliament to enact the Clean Air Act in 1956, which has over time improved the air quality over London significantly.

The United States passed the Air Pollution Control Act in 1955, which was followed by the Clean Air Act of 1963, and subsequent amendments in 1970, 1977, and 1990. Along with the United Kingdom and the United States, most developed countries have passed some form of a Clean Air Act. However, developing countries still face significant challenges. For example, Beijing (China) is well known for air quality problems, as seen in Figure 5.6.

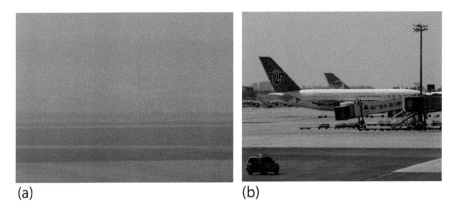

(a)   (b)

FIGURE 5.6 A China Southern A380 on a bad pollution day (left) and on a good pollution day (right) at Beijing Capital International Airport

The World Health Organization considers PM2.5 (particulate matter smaller than 2.5 μm) readings of 25 micrograms per cubic meter as the maximum safe level, but in 2015, Beijing had several instances of readings just under 300 micrograms per cubic meter. This pollution is not only caused by traffic and factory pollution in Beijing, but also by the steel factories and power plants that surround Beijing in Hebei Province. The EPA has also established limits for PM2.5 with the following breakpoints for six air quality index categories:

- Good: $0.0–12.0$ μg/m^3, 24-hour average
- Moderate: $12.1–35.4$ μg/m^3, 24-hour average
- Unhealthy for sensitive groups: $35.5–55.4$ μg/m^3, 24-hour average
- Unhealthy: $55.5–150.4$ μg/m^3, 24-hour average
- Very unhealthy: $150.5–250.4$ μg/m^3, 24-hour average
- Hazardous: >250.5 μg/m^3, 24-hour average

While strides are being made in improving air quality, there is still much work to be done.

The accumulation of pollution is a function of emission rates, dispersion rates, and generation/destruction rates by chemical reaction. Therefore, cities, such as Los Angeles, which sit in a valley, may have lower emission rates than cleaner sites, but because of the surrounding mountains, the pollution is not able to disperse. While highly sophisticated tools have been developed to quantify air quality, modeling of the emissions is still essential for future pollution prediction. Another benefit to developing models is that physical tools are often very expensive (both time and money wise), and models can examine literally an infinite number of scenarios. Finally, modeling is 100% repeatable, which allows the ability to test various scenarios. While accuracy of the models may not be perfect, exact conditions can be repeated.

In order to utilize models for emissions, there are several key dispersion principles that need to be followed. The pollutants generally come from a point source, such as a chimney or smoke stack. The wind will disperse the pollutants both horizontally and vertically into the air. Once in the air, the pollutants encounter both laminar and turbulent flow conditions. These conditions contain eddies and swirls, which are macroscopic random fluctuations from average flow and cause pollution to disperse in potentially unpredictable ways. Eddies form from thermal and mechanical influences. For example, thermal energy from the sun is absorbed into ground, converted to heat, and the heat rises to lowest levels of air by conduction and convection creating thermal eddies. Mechanical eddies, on the other hand, are from shear forces when air flows over rough surfaces. In addition to the eddies, wind fluctuations (speed and direction) also influence pollution dispersion. In order to account for all of these random events, the pollution plume must be considered on time-averaged basis. In general, time averaged distribution is normally distributed, both in the horizontal and vertical directions. Such type of distribution in the horizontal and vertical directions is referred to as binormal distribution of pollutants.

A popular model that quantifies this binormal distribution is the Gaussian Model. The Gaussian Model models dispersion of nonreactive gaseous pollutant from elevated source and is the basis for almost all computer programs developed by environmental protection agencies. The Gaussian Model is shown in Equation 5.6:

$$C = \frac{Q}{2\pi u \sigma_y \sigma_z} \exp\left(-\frac{1}{2}\frac{y^2}{\sigma_y^2}\right)\left[\exp\left(-\frac{1}{2}\frac{(z-H)^2}{\sigma_z^2}\right) + \exp\left(-\frac{1}{2}\frac{(z+H)^2}{\sigma_z^2}\right)\right] \quad (5.6)$$

where

C	= steady-state concentration at a point (x, y, z) ($\mu g/m^3$)
Q	= emissions rate ($\mu g/s$)
σ_y, σ_z	= horizontal and vertical spread parameters (m)
u	= average wind speed at stack height (m/s)
y	= horizontal distance form plume centerline (m)
z	= vertical distance form ground level (m)
H	= h + Δh effective stack height
h	= physical stack height (m)
Δh	= plume rise (m)
σ_y	= ax^b
σ_z	= cx^d + f
x	= horizontal distance from plume origination (km)

and a, b, c, d, and f come from Tables 5.11 and 5.12.

There are some general guidelines when using the Gaussian Model that can help simplify the analysis of pollution dispersion. For example, downwind concentration at any location directly proportional to source strength (Q), and the downwind ground level (z=0) concentration generally inversely proportional to wind speed. In addition, the elevated plume centerline concentrations continuously decline with increasing x, and the ground level centerline concentrations start at zero, increase to a maximum, and then begin decreasing. Finally, the dispersion parameters (σ_y,

TABLE 5.11
Atmospheric Stability Under Various Conditions[a]

Wind speed 10m above ground (m/s)	Daytime solar radiation			Night cloudiness	
	Strong[b]	Moderate[c]	Slight[d]	Cloudy (≥50%)	Clear (<50%)
<2	A	A–B	B	E	F
2–3	A–B	B	C	E	F
3–5	B	B–C	C	D	E
5–6	C	C–D	D	D	D
>6	C	D	D	D	D

[a] Class D applies to heavily overcast skies, any windspeed day or night
[b] Clear summer day with sun higher than 60° above horizon
[c] Summer day with few broken clouds, clear day with sun 35–60° above horizon
[d] Fall afternoon, cloudy summer day, clear summer day with sun 15–35° above horizon

TABLE 5.12
Values of Curve-Fit Constants

Stability	a	b	x < 1.0 km			x > 1.0 km		
			c	d	F	c	d	f
A	213	0.894	440.8	1.941	9.27	459.7	2.094	-9.6
B	156	0.894	106.6	1.149	3.3	108.2	1.098	2.0
C	104	0.894	61.0	0.911	0	61.0	0.911	0
D	68	0.894	33.2	0.725	-1.7	44.5	0.516	-13.0
E	50.5	0.894	22.8	0.648	-1.3	55.4	0.305	-34.0
F	34	0.894	14.35	0.740	-0.35	62.6	0.180	-48.6

σ_z) increase with increasing atmospheric turbulence and the maximum ground-level concentration decreases as effective stack height increases.

Example Problem 5.6

A coal powerplant in Southwest Arkansas is looking to determine the EPA established limit for PM2.5 from its primary emission stack, which is 35 m tall and is emitting 35 µg/s of PM2.5. Specifically, they are looking at the PM2.5 that a local soccer complex would be exposed to that is 1.2 km away from the emission stack. Assume the following conditions: strong daytime solar radiation, a wind speed of 4 m/s 10 m above ground, and the soccer complex is 100 m from the plume centerline.

The first step is calculating σ_y and σ_z. In order to calculate these two values, first the level of stability needs to be established. According to Table 5.11, with the given wind speed of 4 m/s at 10m above ground, and the strong daytime solar radiation, the stability level is B. Next, using Table 5.12, since the distance from the emission stack to the soccer complex is greater than 1.0 km, the following values are used for a-f:

$$a = 156$$
$$b = 0.894$$
$$c = 108.2$$
$$d = 1.098$$
$$e = 2.0$$

With these values established, σ_y and σ_z can be calculated using:

$$\sigma_y = ax^b = (156)\times(1.2)^{0.894} = 183.6 \text{ m}$$

$$\sigma_z = cx^d + f = (108.2)\times(1.2)^{1.098} + 2 = 134.2 \text{ m}$$

Finally, Equation 5.6 can be used to find the steady-state concentration at the soccer complex:

$$C = \frac{Q}{2\pi u \sigma_y \sigma_z} \exp\left(-\frac{1}{2}\frac{y^2}{\sigma_y^2}\right)\left[\exp\left(-\frac{1}{2}\frac{(z-H)^2}{\sigma_z^2}\right)+\exp\left(-\frac{1}{2}\frac{(z+H)^2}{\sigma_z^2}\right)\right]$$

$$C = \frac{35}{2\pi \times 4 \times 183.6 \times 134.2} \exp\left(-\frac{1}{2}\frac{100^2}{183.6^2}\right)\left[\exp\left(-\frac{1}{2}\frac{(35-35)^2}{134.2^2}\right)+\exp\left(-\frac{1}{2}\frac{(35+35)^2}{134.2^2}\right)\right]$$

$$= 3.65^{-4} \text{ µg/m}^3$$

3.65^{-4} µg/m³ falls within the "good" air quality index according to the EPA's established limit.

5.4   COASTAL RESILIENCE – LIVING SHORELINES

According to the Environmental Protection Agency (EPA), coastal areas are particularly vulnerable to the effects of climate change (EPA, 2018). As sea levels rise and frequency of heavy storms increases, coastlines are eroding, resulting in the loss of natural habitat. Coasts are comprised of diverse ecosystems, ranging from beaches, intertidal zones, reefs, seagrasses, salt marshes, estuaries, and deltas (USGCRP, 2018). Loss of these habitats could significantly decrease biodiversity, particularly in coastal areas. Global temperatures are expected to continue to climb, which could lead to saltwater intrusion into coastal aquifers, elevated groundwater tables, changes in local rainfall and river runoff, as well as increased water and surface air temperatures.

In addition to negative effects of climate change, because coastlines are economic hubs for many countries, deterioration of these areas could lead to significant damage to infrastructure as well as population displacement. In 2013, 42% of the United States' population lived in coastal shoreline counties (Kildow *et al.*, 2016). Coastlines provide a uniquely beneficial environment to live and work, with support and resources for jobs in defense, fishing, transportation, and tourism. With these industries, coastal counties were determined to make up nearly half, 48%, of the US gross domestic product (GDP) in 2013. In terms of infrastructure, the Federal Highway Administration (FHWA) estimated more than 60,000 miles of US roads and bridges located in coastal floodplains were at risk to extreme storms and hurricanes which could cause damage costing billions of dollars in repairs (FHWA, 2008).

With the environmental, economic, and social systems which rely so heavily on coastal regions, it is imperative that efforts be taken to remediate and improve the resilience of these areas. Living shorelines, also known as green infrastructure, may help to counter the impacts of environmental changes and protect human development. Living shorelines can help bind sediments, reduce waves, and grow upward as sea levels rise, thus reducing coastline erosion or property damage. In contrast with built infrastructure, or gray infrastructure, such as bulkheads or sea walls, these living shorelines can provide benefits for multiple ecosystems by improving water quality, aquatic habitat, and carbon sequestration. Rather than simply reacting to negative effects of climate change or coastline deterioration, living shorelines can help restore these areas and mitigate environmental issues moving forward.

Living shorelines can be built with only organic materials or a mixture of structural and organic materials, including native wetland plants, stone and rock structures, oyster reefs, submerged aquatic vegetation, coir fiber logs, and sand fill (EPA, 2018). Examples of living shoreline solutions, along with a description and potential applications, are briefly introduced in Table 5.13 and discussed in greater depth in the following paragraphs.

With a position on the front line of many coastal hazards, mangroves have the potential to decrease exposure and vulnerability to a number of risks. Mangroves thrive in many coastal settings, which indicates an ability to cope with hazards, or at a minimum, recover quickly from major impacts. The primary benefits associated with mangroves in terms of coastal resilience are the ability to mitigate the force of waves, reduce storm damage, decrease erosion and bind soil together, with an increase in sea level (Spalding *et al.*, 2014). The complex aerial root systems help slow water flows in a similar manner to energy dissipators which may be used in a flume or at pipe outlets. As high-velocity flows such as waves or storm surges enter the mangrove, water is redirected around an energy dissipator, which is the trees or root system in this case. The roughness of the root system can also help slow wave velocity, much like the use of rough concrete in a flume or rocks placed at pipe outlets. Mangroves have been shown to reduce the energy of waves between 13%

TABLE 5.13

Living Shoreline Applications (Coastal Resilience, 2018)

Green Infrastructure	Description	Application
Mangrove	Group of tree and shrubs which grow in upper inter-tidal zone in warm coastal areas	Reduce wind swell and waves, storm surges, and can rise at similar rates to sea level
Coral Reef	Coral reefs, shellfish reefs, and other physical structures built by corals, molluscs, worms, or algae	Reefs reduce the strength of waves
Oyster Reef	Oysters cluster on older shells, rocks, piers, or other hard, submerged surface; eventually fuse together forming rock-like reefs	Filter suspended particles in water; act as a natural breakwater to stabilize shorelines
Seagrass	Seagrasses are flowering plants which grow entirely underwater	Reduces sediment suspension and creates deep, tightly matted soils which help raise the elevation of the sea bed
Native Wetland Plants	Vegetation growing or planted on the shoreline	Reduce erosion through increased soil stability along coastline

and 66% over 328 ft (100 meters) of mangrove (Spalding *et al.*, 2014). Once waves or storm surges pass through mangroves, the velocity of flow is reduced, making erosion or coastal damage less likely. Additionally, with sea levels continually rising due to climate change, the ability of mangroves to build up soil levels with rising sea level is becoming increasingly beneficial. Active growth of mangroves contributes to increased soil volume through the capture of coastal sediments, allowing the tree height itself to keep pace with sea-level changes.

Coral reefs and oyster reefs can also act as a buffer to protect coastlines from the negative impacts of waves and storm surges. Comprised of coral, oysters, shellfish, and other hard physical structures built by corals and mollusks, reefs can act as a natural breakwater to dampen the force of waves driving toward coastlines (Coastal Resilience, 2018). Similar to mangroves, reefs serve to dissipate and decrease the energy of waves by redirecting and slowing the velocity. Geometry of reefs influences the ways in which wave energy is attenuated. For instance, living coral is more shallow and rougher with more complex structures, dissipating energy of waves primarily through friction. As waves encounter the living coral, the velocity of the water is opposed with frictional forces due to surface roughness of the coral, much like the use of angular aggregate along a ditch near a stormwater pipe outlet.

A second and unique benefit to oyster reefs is the ability to filter suspended particles in the water. Oysters feed on algae sifted from the water. As algae are filtered out, water quality is improved through the removal of excess nutrients. In one day, an oyster can filter up to fifty gallons of water per day. One study which sought to take in-situ measurements of suspended solids uptake to quantify water quality improvements associated with oyster reefs found uptake rates ranging from 27.8% to 62.3% (Grizzle *et al.*, 2006). This range is due to differences in species, oyster size, oyster density, water flow rate, and water depth. Similar variables also influence the removal rates of traditional filtration systems, such as sand filters, which consider depth of filter media, filtration rate, and filter media particle size.

Seagrass, which grows in large meadows in subtidal ecosystems, can also provide an additional line of defense for coastlines by reducing suspended sediment, raising the elevation of the seabed and reducing erosion. Unlike oysters, which filter water through and pull out suspended particles through a feeding process, seagrass collects suspended particles within the grass as waves pass through the meadows, much like mangroves. Also like mangroves, trapped sediment is deposited on the seabed, bound together by the root system, and allows the plant height to continue to grow as the seabed rises. Finally, reducing erosion near the shoreline is a major concern for coastlines which can be addressed using seagrass. Because the grass can dissipate wave energy, erosion is indirectly decreased through the reduction in forces of the waves. The seagrass root system is also believed to increase the shear resistance of soils below the seagrass, stabilizing the sediment.

Each of the previously mentioned living shoreline strategies have provided defense off the coast; however, native wetland plants can provide coastal protection on the shoreline itself by increasing slope stability. The use of plants and vegetation leads to multiple beneficial effects which can improve slope stability (Clark, 2005). First, moisture remaining in the slope soil is decreased due to leaves intercepting rainfall and the root system extracting moisture for the plant itself. The roots also serve to reinforce the soil by adding shear strength. As roots grow deeper into the slope, structure is provided to support soils upslope of the roots, forming slope layers. These layers somewhat mimic the use of geogrids on slopes. When quantifying slope stability, the factor of safety against slope instability, which is calculated using Equation 5.7, is often used.

$$FS = \frac{T_{FF}}{T_{MOB}} \tag{5.7}$$

where T_{FF} represents the available shearing resistance and T_{MOB} is the mobilized shear force along the slip surface. Using native plants could increase shear parameters, such as cohesion, c, and friction angle, ϕ, which would increase shearing resistance and increase the factor of safety.

It should be mentioned, living shorelines can be used in conjunction with traditional, gray infrastructure, such as sea walls and jetties. The US Army Corps of Engineers refers to these as Integrated Coastal Risk Reduction Approaches. A successful example of a solution which integrated green infrastructure and gray infrastructure is the Mississippi Coastal Improvement Plan implemented after Hurricane Katrina to address hurricane and storm damage reduction, salt water intrusion, shoreline erosion, and wildlife preservation (USACE, 2009). Both the engineering attributes of components as well as the interactions between green and gray infrastructure should be considered over the short and long terms. Structural, gray infrastructure could potentially weaken natural defenses provided by green infrastructure.

Instead, an integrated approach should seek to build multiple lines of defense, balancing both structural and natural solutions (Bridges *et al.*, 2013).

Coastal Resilience, an organization of the Nature Conservancy dedicated to helping communities understand vulnerability and reducing risks of coastal hazards, has developed a four-step approach to utilizing nature-based solutions to improve resilience of coastal areas (Coastal Resilience, 2018):

1. Assess Risk and Vulnerability
2. Identify Solutions
3. Take Action
4. Measure Effectiveness

Risk assessment is an integral part of this approach as it identifies who or what is at risk and the characteristics which make the ecological system more or less vulnerable. Coastal Resilience defines a risk assessment framework with the relationships shown in Equations 5.8 and 5.9:

$$\text{Risk} = \text{Exposure} \times \text{Vulnerability} \qquad (5.8)$$

$$\text{Vulnerability} = \text{Susceptibility} \times \text{Coping Capacity} \times \text{Adaptive Capacity} \qquad (5.9)$$

In these relationships, exposure refers to those identified as being at risk. Vulnerability then describes aspects or circumstances of a system which make it more susceptible to damage.

Once risk and vulnerability have been quantified, solutions should be identified. As civil engineers, analyzing alternatives by considering not only the engineering problem but also environmental, economic, and social implications is necessary. Living shoreline solutions will require collaborative efforts with professionals outside the field of civil engineering with expertise in these types of ecosystems. Factors such as climate, existing infrastructure, and risks to be addressed should be considered when selecting appropriate solutions. Objectives of a solution may include protecting or restoring coast, linking natural and built defense structures, removing incentives to build in high-risk areas, and meeting specific needs of the community.

The third step, while seemingly simple, refers not only to implementing solutions but being willing to pursue sustainable and innovative solutions. Traditionally, adaptation to hazards along coastlines has involved shoreline hardening and engineered defenses, rather than a green infrastructure, ecosystem-based approach. While traditional approaches have previous performance and engineering design supporting use, living shorelines provide a natural solution and unique benefits which restore health of degraded habitats and improve resilience of the area. Organizations such as the Nature Conservancy and the National Oceanic and Atmospheric Administration have partnered to develop a number of tools and resources to educate communities and decision makers on green infrastructure alternatives, and even worked with communities to implement such solutions. Taking action requires engineers who are

willing to join these efforts, be creative, and look beyond traditional design to identify solutions which improve quality of life for future generations.

Finally, after a solution has been implemented, the effectiveness should be monitored. In order to determine if a solution is successful and identify areas of future
improvement, relevant performance measures should be developed and tracked. These
performance measures should be related to the risk and vulnerabilities identified in the
first step, understanding how the solution has mitigated these issues and improved resilience. Performance measures may also be related to educating others on the viability of
living shorelines as a solution to issues facing coastal regions. Examples of these performance measures may include publications, collaborative outreach efforts developed,
individuals trained, or even increased stakeholder awareness.

Example Problem 5.7

An area of coastline along Lake Michigan is at risk of continuing to erode, which
is expected to cause damage to houses built near the coast, and the use of native
wetland plants has been proposed to increase the slope stability. Calculate the
expected factor of safety against slope instability along this area of coastline after
native plants are used for remediation. The predicted slip surface would have a
length of 10 ft at a 30° angle. For the analysis area, this slip surface would displace
about 20 ft³ per linear foot of soil along the coast. Assume the density of soil is 130
lb/ft³, cohesion is 250 psf, and the internal friction angle is 45° after the addition of
native plants. The factor of safety before remediation is 1.2. Can you recommend
the use of native plants as a viable option for improving slope stability and successful coastal remediation in this case?

To begin calculating the available shearing resistance along the slip surface, the
weight of soil above the slip surface should be calculated:

$$W_M = \frac{\text{Density}}{\text{Volume}}$$

$$W_M = \frac{130\dfrac{\text{lb}}{\text{ft}^3}}{20\text{ft}^3} = 2600\text{lb}$$

Now, the mobilized shear force along the slip surface, T_{MOB}, can be calculated
as follows:

$$T_{MOB} = W_M \sin\alpha_S$$

$$T_{MOB} = 2600\text{lb}\times\sin(30°) = 1300$$

The available shearing resistance along slip surface is then calculated:

$$T_{FF} = cL_S + W_M \cos\alpha_S \tan\phi$$

$$T_{FF} = 250\frac{\text{lb}}{\text{ft}^2}\times10\text{ft} + 2600\text{ lb}\times\cos(30°)\times\tan(45°) = 4751.7$$

Therefore, the final factor of safety against slope instability after remediation using native plants would be determined to be:

$$FS = \frac{T_{FF}}{T_{MOB}} = \frac{4751.7}{1300} = 3.66$$

Because the factor of safety after remediation is nearly three times that of what it was before, you could confidently recommend this alternative as a viable solution.

SIDEBAR 01
Constructing a High-Quality Graph

A well-constructed graph must be easy to read and understand quickly and should be constructed so it can be read both in color and in black and white. The structure of the graph must include both horizontal (x-axis) and vertical (y-axis) axis titles and a legend. If the graph is within a document, it should not have an embedded chart title, as the caption provides the necessary information. However, if the graph is standalone a chart title can be used. The axis titles and chart title should be bolded, and one font size larger than the axis labels. The legend font should be the same size as the axis label font and is not bolded. The axis labels should be large enough where you only have 4–6 delineations on the y-axis. Graphs are not used to convey precise numbers, but to show trends. With too many delineations, the axis become cluttered and unreadable. Within the delineations, the number of significant digits should remain the same, and the last digits should not all be zeros. If they are all zeros, remove significant digits until the last digits are not all zero, unless you are to the left of the decimal. Finally, rarely are all data points shown in a graph, especially one showing multiple data points per second over a time period of minutes. By only showing a portion of the data, the plotted points can be connected, which allows for clearer data presentation.

Let's take a look at calculating fracture energy. Fracture energy is the amount of energy required to create a new surface in a material. This is done in the lab by either pulling on, or pushing on, a sample to create a crack. The information recorded during this test is the load and some sort of displacement. The area under this load/displacement curve (the work), divided by the area of the cracked surface created, is the fracture energy. However, it is often interesting to look at the load/displacement curves, and to show how they are different relatively to one another. For example, asphalt mixtures are viscoelastic materials, meaning at higher temperatures they show more viscous behavior, and at lower temperatures they show more elastic behavior. Therefore, it would be expected that the work required to create a crack surface would be higher at the higher temperatures, and lower at the lower temperatures. This trend is shown in Figure S5.1. Figure S5.1a has the standard output from Microsoft Excel, whereas Figure S5.1b is modified according to the "rules of

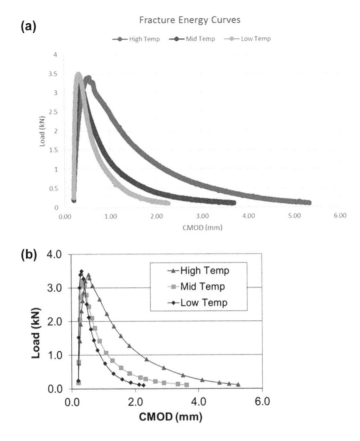

FIGURE S5.1   a) Standard Excel output of fracture energy curves; b) modified Excel output of the same fracture energy curves

thumb" outlined in this sidebar. It is obvious that by taking some care, graphs can become much easier to read and interpret, especially if shown in black and white.

HOMEWORK PROBLEMS

1. You have just graduated from college (congratulations!) and on the first day of work, your boss asks you to explore the variability of runoff coefficients on peak discharge. Assuming a rainfall intensity of 12.7 mm/hr and a watershed are of 404,686 m2, graph the minimum and maximum potential runoff for forest, lawn with clay soil, and a slope between 2% and 7%, and asphalt. When establishing minimum and maximum potential runoffs, utilize the PE Reference Manual, the City of Fayetteville, Washington State,

and Florida State. Briefly describe the trends that you see, and use the format provided under the sidebar in Chapter 5 "Constructing a High-Quality Graph."

2. Find a building with a flat roof nearby and estimate the area of the roof. Assume the roof has shingles, and calculate the peak discharge of the roof currently, and how it will change if replaced by a green roof. Use a rainfall intensity of 0.75 in/hr and state any assumptions you need to make.

3. The drinking water district of Salt Lake City, UT, wants to compare bark, sugar beet pulp, and exhausted coffee to the traditional granular-activated carbon for removing chlorite from their drinking water. Assuming the Q_{max} values in Table 5.9 can be used for chlorite, and that $b = 0.237$ for all four materials, how much material must be used in order to treat one liter of water to meet the EPA maximum requirement? How much material must be used in order to treat one liter of water to meet the EPA maximum goal?

4. Plot the change in maximum sorption capacity of sunflower stalks for metal ions contaminant, the size, and the temperature. Try to fit all of the data onto a single, easy-to-read graph, using the format provided under the sidebar in Chapter 5 "Constructing a High-Quality Graph."

5. A coal power plant in Champaign, Illinois, is looking to determine the EPA established limit for PM2.5 from its primary emission stack, which is 15 m tall and is emitting 412 μg/s of PM2.5. Specifically, they are looking at the PM2.5 that a local softball complex would be exposed to that is 0.3 km away from the emission stack. Assume the following conditions: nighttime, cloudy, a windspeed 5.5 m/s 10 m above ground, and the softball complex is 50 m from the plume centerline.

6. Using the data in homework problem 7, what is the minimum emission of PM2.5 (in μg/s) that would be required to fall into the "moderate" air quality index category established by the EPA?

7. Suppose you are responsible for making a recommendation for improving coastal resilience along the coastline near New Orleans, Louisiana. The two green infrastructure alternatives being considered are mangroves and oyster reefs. Using the format provided under the sidebar in Chapter 1 "Writing a High-Quality Essay," compare the benefits of selecting mangroves to those of selecting oyster reefs. Provide two discussion points for each alternative.

8. Native plants are being considered for improvement of slope stability along the coastline in northern Oregon. The predicted slip surface is 15 ft long at an angle of 20°. The slip surface is anticipated to displace 30 ft^3 per linear foot of soil. After native plants are placed along the coastline, soil would have a density equal to 127 lb/ft^3, cohesion value of 255 psf, and internal friction angle of 45°. If the factor of safety against slope instability before remediation is equal to 1.9, are native plants a viable choice for improving slope stability in this case?

REFERENCES

Bean, E., Hunt, W., Bidelspach, D. Evaluation of Four Permeable Pavement Sites in Eastern North Carolina for Runoff Reduction and Water Quality Impacts. *Journal of Irrigation and Drainage Engineering*, December 1, 2007, 133(6), 583–592.

Bridges, T., Henn, R., Komlos, S., Scerno, D., Wamsley, T., White, K. Coastal Risk Reduction and Resilience. U.S. Army Corps of Engineers, 2013.

Clark, G. Stabilizing Coastal Slopes on the Great Lakes. University of Wisconsin Sea Grant Institute, 2005.

Coastal Resiliency. EPA, Environmental Protection Agency, 2018. www.epa.gov/green-infrast ructure/coastal-resiliency.

Coastal Resilience. Natural Solutions. 2018. coastalresilience.org/natural-solutions/.

EPA. Storm Water Technology Fact Sheet: Bioretention. United States Environmental Protection Agency, EPA 832-F-99-012, September 1999.

EPA. National Primary Drinking Water Regulations. United States Environmental Protection Agency, EPA 816-F-09-004, May 2009.

Fassman, E., Blackbourn, S. Urban Runoff Mitigation by a Permeable Pavement System Over Impermeable Soils. *Journal of Hydrologic Engineering*, June 1, 2010, 15(6), 475–485.

FHWA. *Highways in the Coastal Environment*, Second Edition. Hydraulic Engineering Circular No. 25. FHWA-NHI-07-096. Federal Highway Administration. Department of Civil Engineering, University of South Alabama, Mobile, AL, 250 pp, 2008.

FHWA. Our Nation's Highways. U.S. Department of Transportation, Federal Highway Administration, 2011.

Grizzle, R. E., Greene, J. K., Luckenbach, M. W., L. D. Coen. New In Situ Method for Measuring Seston Uptake By Suspension-Feeding Bivalve Molluscs. *Journal of Shellfish Research*, 25(2), pp. 643-649, 2006.

GSA. The Benefits and Challenges of Green Roofs on Public and Commercial Buildings. A Report of the United States General Services Administration, May, 2011.

Hansen, K. Porous Asphalt Pavements for Stormwater Management. IS 131, National Asphalt Pavement Association, Lanham, MD, 2008.

Kildow, J. T., C. S. Colgan, P. Johnston, J. D. Scorse, and M. G. Farnum. State of the U.S. Ocean and Coastal Economies: 2016 Update. National Ocean Economics Program, Monterey, CA, 31 pp, 2016.

Kumar, U., Bandyopadhyay, M. Sorption of Cadmium from Aqueous Solution Using Pretreated Rice Husk. *Bioresource Technology*, 2006, 97, 104–109.

Mobilia, M., Longobardi, A., Sartor, J. Impact of Green Roofs on Stormwater Runoff Coefficients in a Mediterranean Urban Environment. Recent Advances in Urban Planning, Sustainable Development and Green Energy. *Proceedings of the 5th International Conference on Urban Sustainability, Cultural Sustainability, Green Development, Green Structures and Clean Cars*, USCUDAR, Florence, Italy, November 22–24, 2014, pp. 100–106.

Moran, A., Hunt, W., Smith, J. Green Roof Hydrologic and Water Quality Performance from Two Field Sites in North Carolina. *Managing Watersheds for Human and Natural Impacts: Engineering, Ecological, and Economic Challenges, Watershed Management Conference 2005*, Williamsburg, VA, United States, July 19–22, 2005, pp. 1–12.

NYC. Guidelines for the Design and Construction of Stormwater Management Systems, New York City Department of Environmental Protection, July, 2012.

Spalding M, McIvor A, Tonneijck FH, Tol S, van Eijk P. Mangroves for Coastal Defense. Guidelines for Coastal Managers & Policy Makers. Published by Wetlands International and The Nature Conservancy. 42 pp., 2014.

St. John, M. Effect of Road Shoulder Treatments on Highway Runoff Quality and Quantity. M.S. Thesis, University of Washington, 1997.

Sun, G., Shi, W. Sunflower Stalks as Adsorbents for the Removal of Metal Ions from Wastewater. *Industrial Engineering Chemical Research*, 1998, 37, 1324–1328.

USACE. Mississippi Coastal Improvement Plan (MsCIP). Mobile: USACE Mobile District, 2009.

USGCRP. Impacts, Risks, and Adaptation in the United States: Fourth National Climate Assessment, Volume II [Reidmiller, D.R., C.W. Avery, D.R. Easterling, K.E. Kunkel, K.L.M. Lewis, T.K. Maycock, and B.C. Stewart (eds.)]. U.S. Global Change Research Program, Washington, DC, USA, 1515 pp, 2018.

Wei, I. Installation and Evaluation of Permeable Pavement at Walden Pond State Reservation. Project 77–12 and 80–22, Division of Water Pollution Control, The Commonwealth of Massachusetts, February, 1986.

WHO. Technical Notes on Drinking-Water, Sanitation, and Hygiene in Emergencies. World Health Organization, Water, Sanitation, Hygiene and Health Unit, July 2013.

6 Application
Geotechnical Sustainability

> The function of education is to teach one to think intensively and to think critically. Intelligence plus character – that is the goal of true education.
>
> **Martin Luther King Jr**

As many geotechnical engineers like to say, soils are the foundation of everything. In a sense this is true, because the vast majority of engineering structures are built on land, which is generally composed of some sort of soil. Therefore, a strong understanding of soil is absolutely necessary in order to design the structure, roadway, or any artificial body of water that rests on the soil because regardless of what you are designing or construction, it will almost always eventually involve the ground. This understanding includes the plasticity and structure, the compaction and permeability characteristics, and the stresses and pressures that move through soil. When considering sustainability application in geotechnical areas, there are two primary paths, either the replacement of materials or the modification of a design in order to improve the economic, environmental, and social impact of the project. This chapter will cover both of these paths using the following four specific applications:

1. Alternate granular fill materials
2. Retaining wall design
3. Mechanically stabilized earth walls
4. Geothermal energy foundations

6.1 ALTERNATE GRANULAR FILL MATERIALS

The natural topography of a site is rarely appropriate for an engineering structure without first modifying the site grading in some manner. During the construction of highways, railways, buildings, landfills, dams, or levees, materials often need to be either removed (commonly known as "cut") or added (commonly known as "fill") in order to properly prepare a site. There are three typical formulas used to calculate the amount or volume (V) of cut and fill of a site: the average end area formula, the prismoidal formula, and the pyramid formula. The average end area formula simply takes the area of one end of the cut or fill (A_1), adds this area to the opposite end (A_2), multiples by the length of the cut or fill (L), and divides by two, as shown in Equation 6.1.

$$V = \frac{L(A_1 + A_2)}{2} \tag{6.1}$$

The prismoidal formula uses a similar calculation, but includes the area of the mid-section of the cut or fill (A_m), creating a slightly more accurate estimation of the volume of material necessary, as shown in Equation 6.2.

$$V = \frac{L\left(A_1 + 4A_m + A_2\right)}{6} \tag{6.2}$$

The third formula, the pyramid or cone, is used if the site tapers to a single point with a known height (h) and area of base (A_b), as shown in Equation 6.3.

$$V = \frac{h\left(A_b\right)}{3} \tag{6.3}$$

All of the areas can be calculated by using various methods, including the area by coordinates, the trapezoidal rule, or Simpson's 1/3 rule, all of which were learned in Calculus I. However, it is often necessary to switch from known volumes to known weights when dealing with cut and fill material. There are two reasons for this. First, properties of the cut and fill material are highly dependent on the amount of water in the material. This water can be found either as a discrete phase (free water) or within the voids, so the knowledge of volume alone is often not adequate for a full understanding of the material. Second, it is very difficult to purchase or sell cut and fill material on the basis of volume. The transaction is usually conducted by weight, as trucks can easily pull on and off scales, but are rarely measured by height, width, and length, to establish quantities. Therefore, understanding phase relationships, between weight and volume, is critical. This can be done using a phase diagram, as seen in Figure 6.1

The key to moving between the weight of a material and the volume of the material is the specific gravity (γ) and the moisture content (w). Beginning with the specific gravity of water ($\gamma_w = 62.4$ lb/ft^3 or 9.81 kN/m^3), and incorporating components of the phase diagram, various relationships can be established, including the specific gravity of the solids (G_s), given in Equation 6.4.

$$G_s = \frac{\left(W_s/V_s\right)}{\gamma_w} \tag{6.4}$$

The void ratio (e) in Equation 6.5:

$$e = \frac{V_v}{V_s} \tag{6.5}$$

The saturated unit weight (γ_{sat}) in Equation 6.6:

$$\gamma_{sat} = \frac{\left(G_s + e\right)\gamma_w}{\left(1 + e\right)} \tag{6.6}$$

And finally, the porosity (n) in Equation 6.7:

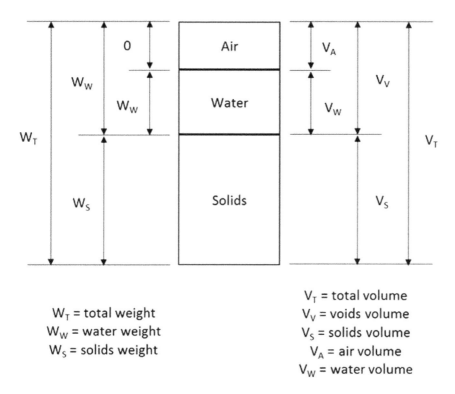

FIGURE 6.1   Phase diagram of cut or fill material

$$n = \frac{V_v}{V} = \frac{e}{(1+e)} \tag{6.7}$$

These relationships, among others, allow for cut and fill to be fully evaluated and categorized in the laboratory to ensure proper behavior in the field.

Embankment refers to placing and compacting materials to raise the existing grade above the level of surrounding ground surface, usually for a roadway, a railway, or the area under a building pad. Fill refers to placing and compacting material in a depression or hole, or the leveling of an existing site for preparation of a slab on grade foundation. When considering fill under a pavement structure, generally coarser and lower-quality material is placed at the bottom to provide a firm foundation and drainage, while the top portions are well-compacted with high-quality material that can support the structure being constructed directly above. However, fill under a building foundation is not as straight forward. Depending on the depth of the fill, the lower materials may need to be just as highly controlled as the upper materials or the foundation bearing capacity and settlement design will be compromised. Important properties of alternate granular fill materials include gradation, unit weight/specific gravity, moisture-density characteristics (optimal moisture content, maximum dry density), shear strength (cohesion, internal friction), and compressibility (consolidation, settlement). Of these properties, the gradation is the only property that can be influenced, by crushing, screening, or washing.

Traditionally, cut and fill materials consist of natural soils. Ideally, cut materials from one portion of the project are utilized as fill materials in another part of the project (on-site borrow), which minimizes haul distance, thereby reducing cost (economic pillar of sustainability) and emissions (environmental pillar of sustainability). However, another sustainable alternative is to use alternate granular fill materials, or by-product materials. This is not only potential good from an economic and environmental standpoint, but bringing in alternate material may be necessary if the existing material is not suitable for fill, such as if the area was formally a landfill or the existing soil is simply of low quality. There are seven relatively common alternate granular fill materials that are used as either cut or fill material: blast furnace slag, coal fly ash, mineral processing wastes, nonferrous slags, reclaimed asphalt pavement, reclaimed concrete material, and scrap tires (Chesner *et al.*, 1998). Table 6.1 summarizes properties of the seven alternate granular fill materials.

Table 6.2 shows similar properties for a sampling of the unified soil classification system (UCSC) soils. Table 6.2 was constructed from various resources, including textbooks and online information. While the trends are consistent (i.e. the optimal moisture content generally increases as you move down the table, while the compacted density, internal friction angle, and CBR generally decrease), the actual data is provided as a general reference, and there may be exceptions.

When considering the use of alternative fill materials, there are three additional considerations that need special attention, including material process requirements, design considerations, and construction procedures. For example, special material process requirements for reclaimed concrete or scrap tires include removing any reinforcing steel. Both of these materials often utilize steel (concrete for tensile strength and tires for sidewall stiffness), but steel is not a desirable material for use as

---

TABLE 6.1

Properties of Alternate Granular Fill Materials (Chesner *et al.*, 1998)

	Optimum moisture content (%)	Compacted density lb/ft³,(kg/m³)	Internal friction angle (°)	CBR (%)
Blast furnace slag	9–19	70–120 (1120–1940)	40–45	>100
Coal fly ash	20–35	85–100 (1380–1600)	26–40	Increase soils ~20x
Nonferrous slags	4–8	175–237 (2800–3800)	40–53	>100
Recycled asphalt pavement	3–7	100–125 (1600–2000)	37	20–25
Reclaimed concrete	4–11	120–180 (1940–2900)	>40	90–140
Scrap tires	1	20–45 (320–720)	19–25	Increase soils ~10x

TABLE 6.2
Properties of Select USCS Fill Materials

USCS Classification (ASTM 2487)	Optimum moisture content (%)	Compacted density lb/ft³, (kg/m³)	internal friction angle (°)	CBR (%)
GW – well graded gravel	8–11	125–135 (2000–2160)	33–40	60–80
GM – silty gravel	8–12	120–135 (1920–2160)	30–40	40–80
SW – well graded sand	9–16	110–130 (1760–2080)	33–43	20–40
SC – clayey sand	10–19	105–125 (1680–2000)	30–40	10–20
CL – lean clay	12–24	95–120 (1520–1920)	27–35	5–15
MH – elastic silt	24–40	70–95 (1120–1520)	23–33	4–8

fill material because of its high weight and the potential to damage equipment during transport, placement, and compaction. Another example, for design considerations, is coal fly ash. While it is similar in many respects to earthen backfill, it tends to wick water to itself which reduces shear strength. Therefore, it often has to be delivered to the job site at the near-optimal moisture content to prevent the absorption of water. Finally, for construction procedures, an example of special consideration again comes from scrap tires, that should be wrapped with non-woven geotextile fabric once the tires are placed and compacted. This will prevent expansion after compaction of the material, which will influence the volume calculations that were used during the design. Care must be taken with using all recycled material, however, ensuring that there are no existing environmental regulations that prohibit the use of any specific material.

Example Problem 6.1

Construct a phase diagram for recycled asphalt pavement, which generally has a specific gravity of 2.565. Use standard units; assume a soil volume of 1 ft³ and a total volume of 1.2 ft³. State any other assumption that you need to make.

The first assumption is for the moisture content. Using Table 6.1, the average optimal moisture content is assumed as the moisture content, so $w = 5.0\%$. We know that $V_s = 1$, so we can calculate the weight of solids using Equation 6.4:

$$G_s = \frac{(W_s/V_s)}{\gamma_w} \rightarrow 2.565 = \frac{(W_s/1)}{62.4 \text{ pcf}} \rightarrow W_s = 160.1 \text{ pcf}$$

Next, the weight of the water can be calculated using:

$$W_w = w \times G_s \times \gamma_w = 0.05*2.565*62.4 \text{ pcf} = 8.0 \text{ pcf}$$

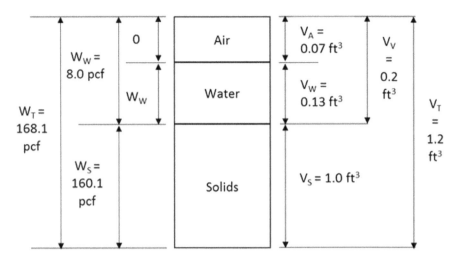

FIGURE EP6.1   Phase diagram for recycled asphalt pavement

The total weight is simply the combination of the solid weight and water weight, or 168.1 pcf. Now that the right side of the phase diagram is complete, the left side can be analyzed. First, the volume of water can also be calculated:

$$V_W = w \times G_s = 0.05 * 2.565 = 0.13 \text{ ft}^3$$

Next, the void volume can be calculated:

$$V_V = V_T - V_S = 1.2 \text{ ft}^3 - 1.0 \text{ ft}^3 = 0.2 \text{ ft}^3$$

And finally, the volume of air can be calculated, giving us the full phase diagram below (Figure EP6.1).

$$V_A = V_V - V_W = 0.2 \text{ ft}^3 - 0.13 \text{ ft}^3 = 0.07 \text{ ft}^3$$

6.2   RETAINING WALL DESIGN

While some job sites may be located in southern Florida, where stating that the land is flat is an understatement, many areas will be located in areas with elevation change. Whether designing a roadway or a structure, the topography may have to be modified, and the change in topography is frequently increased in order to create level areas for the engineering structure. A typical solution is to construct a concrete retaining wall, as shown in Figure 6.2a. However, an alternate solution can be a bioengineered slope, which uses vegetation instead of concrete in order to restrain

(a)   (b)

FIGURE 6.2   a) A traditional concrete retaining wall on the University of Arkansas campus (credit: A Braham). b) A bioengineered slope just north of Milwaukee, WI (credit: A Braham)

the soil, as can be seen in Figure 6.2b. A brief review of earth pressure is provided, followed by some examples of the sustainability of bioengineered slopes.

Rigid retaining walls are generally large masses of vertically constructed concrete or blocks that prevent the earth from moving laterally. A retaining wall without any steel reinforcement is a mass gravity retaining wall, and can be constructed out of plain concrete or stone masonry. As the name implies, their self-weight and any soil resting on the structure stabilize the structure. Lateral sliding of the wall is prevented by the friction between the base of the wall and the bearing soil (whether natural soil or fill material). This type of construction is only economical for wall heights that are no greater than 1.5 times the width of the footing. An example of three general stages of mass gravity wall construction is shown in Figure 6.3.

However, to either increase the height of the wall, or to reduce the cross section of the concrete, steel reinforcements can be used. A steel-reinforced concrete wall is called a semi-gravity retaining wall. Instead of relying on the self-weight of the wall itself, semi-gravity walls utilize the steel reinforcement to resist bending and shear. However, from a design perspective, it is usually cheaper to excavate or construct a slightly bigger wall versus installing steel reinforcement. Nonetheless, as seen in Figure 6.4, by placing steel toward the earth edge of the wall, the thickness of the stem can be reduced, thus reducing the quantity of concrete.

A cantilever retaining wall is also made of reinforced concrete, but consists of a thin stem and base foundation. The portion of the footing in front of the wall surface is known as the toe; the portion of the foundation behind the wall and covered with backfill is known as the heel. The names comes from the cantilever action of the stem retaining the soil mass behind the wall. The weight of the soil on the heel as well as the wall's self-weight assists with achieving wall stability. The shape of a cantilever retaining wall is usually either a T-shape or an L-shape, as seen in Figure 6.5.

Finally, a counterfort retaining wall is similar to a cantilever wall, but also include thin, vertical, concrete slabs (known as counterfort) that tie the wall and

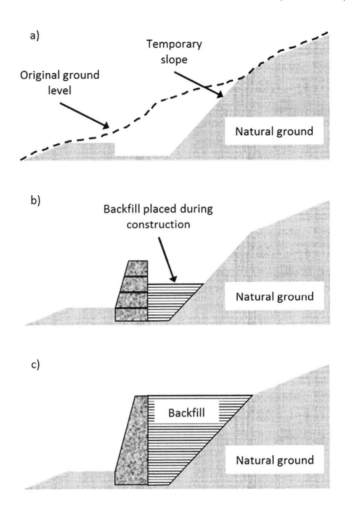

FIGURE 6.3 Simplified construction of a gravity wall: a) excavation, b) wall construction, c) final product (credit: A Braham)

base slab together. These counterforts reduce the shear and bending moments in the wall. Counterfort retaining walls are almost exclusively used for very tall walls, usually 30–36 ft (10–12 m) tall. While counterfort retaining walls are not as common as the other three, it is still an option during design. A simplified wall is shown in Figure 6.6.

Regardless of which of the four walls are used, all are designed to hold the earth back from an engineered area. When holding the earth back, it is necessary to calculate the lateral earth pressure, or how much the soil is being pushed to the wall.

The first case of interest is when the soil is partially saturated. When the soil is partially saturated, an effective horizontal force is created. This effective horizontal

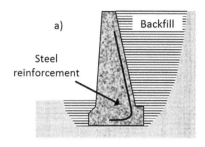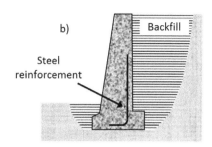

FIGURE 6.4 Typical semi-gravity concrete walls with steel reinforcement (credit: A Braham)

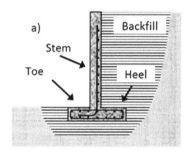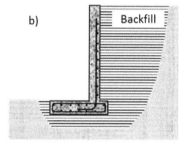

FIGURE 6.5 A T-shape (a) and L-shape (b) cantilever retaining wall (credit: A Braham)

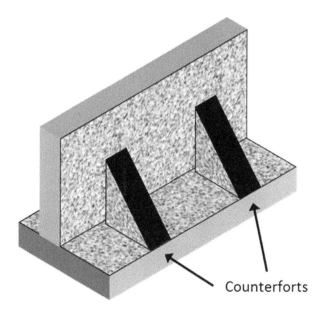

FIGURE 6.6 A simplified counterfort retaining wall (credit: A Braham)

force is a combination of the effective vertical stress and the pore water pressure. The effective horizontal force (P_o) can be solved using Equation 6.8.

$$P_o = \left(P_{A1}\right) + \left(P_{A2}\right) = \left(\frac{1}{2}\sigma_1' K_o H_1\right) + \left(\frac{1}{2}\left(\sigma_1' + \sigma_2'\right)K_o H_2\right) \qquad (6.8)$$

where:

P_o = force per unit length of wall, or effective horizontal force

P_{A1} = force per unit length of unsaturated earth

P_{A2} = force per unit length of saturated earth

σ_1' = lateral pressure of unsaturated earth

  = unit weight of unsaturated earth (γ_1) multiplied by height of unsaturated earth (H_1)

K_o = earth pressure coefficient at rest = $1 - \sin \varphi'$ for coarse grained soils

σ_2' = lateral pressure of saturated earth

  = $(\gamma_1 \times H_1) + (\gamma_2 \times H_2) - (\gamma_w \times H_2)$

γ_2 = unit weight of saturated earth

γ_w = unit weight of water

H_2 = height of saturated earth

The general case of the earth pressure coefficient at rest (K_o) is the ratio of the horizontal effective stress and the vertical effective stress. However, various other functions for coarse grained soils, loose sand, compacted sand, and clays. In addition, work has been done for over consolidation conditions as well.

The calculation of the effective horizontal force is shown graphically in Figure 6.7.

A specific case for the earth pressure coefficient is when every point in a soil mass is about to fail, which is the plastic equilibrium in soil. This was explored by Rankine, who looks at both the active and passive forces on a retaining wall (Rankine, 1857). The active force condition, or K_A, is when the retaining wall is allowed to *move away* from the retained soil mass. Conversely, the passive force condition, or K_P, is when the retaining wall *is pushed into* the soil mass. These conditions are represented by Equations 6.9 and 6.10.

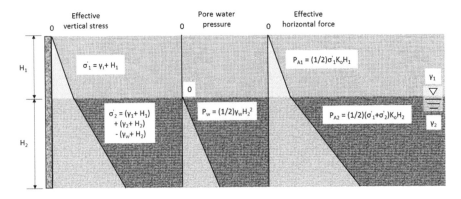

FIGURE 6.7 Horizontal stress profiles and forces (credit: A Braham)

$$K_A = \tan^2\left(45° - \phi/2\right) \tag{6.9}$$

$$K_P = \tan^2\left(45° + \phi/2\right) \tag{6.10}$$

where a smooth wall is assumed, the backfill is level, the cohesion (c) of the soil is zero, and ϕ is the angle of internal friction.

From a sustainability standpoint, there have been several studies that have compared a traditional concrete retaining wall versus a bioengineered slope. For example, Storesund *et al.* examined the Life Cycle Cost Analysis (LCCA), energy consumption, and Global Warming Potential during the planning, design, construction, and operation and maintenance of a creek restoration site (2008). The following life cycle components were included for each stage:

- Planning
  - Permits, permit preparation, configuration and layout, cost and schedule, environmental impact, site-characterization: concrete and bioengineered
- Design
  - Scour evaluation, design analysis, plans/specifications/schedule, material quantities, stormwater runoff: concrete and bioengineered
- Construction
  - Earthwork, formwork, steel and concrete, backfill: concrete only
  - Earthwork, vegetation implementation, erosion control: bioengineered only
- Operation and maintenance
  - Graffiti removal: concrete only
  - Pruning and weeding, vegetation replacement, insect and disease control: bioengineered only

A summary of the analysis is provided in Table 6.3.

In Table 6.3, the actual cost of the bioengineered slope is actually less than the reinforced concrete wall, but because of the higher operation and maintenance cost, the LCCA is actually higher. However, it is important to realize the assumptions that went into the operation and maintenance of the reinforced concrete wall. It was assumed that there would be no deterioration of the concrete, which is certainly possible, assuming proper design and construction. However, if deterioration were included, the LCCA results may have been different. While the economic pillar is higher for the bioengineered slope, the environmental pillar is much lower. The energy of the bioengineered slope is almost 1/3 of the reinforced concrete, and the GWP is just under ½. Therefore, a decision must be made by the owner: is the economic pillar or the environmental pillar more important?

While reinforced concrete walls and bioengineered slopes are two options for designing a retaining wall, a third option exists, a mechanical stabilized earth wall (MSE).

TABLE 6.3
Summary of Reinforced Concrete Versus Bioengineered Retaining Wall

Life cycle stage	Reinforced Concrete			Bioengineered		
	Value ($)	Energy, ft-lb (GJ)	Global Warming Potential, lb (kg)	Value ($)	Energy, ft-lb (GJ)	Global Warming Potential, lb (kg)
Planning	50,300	8.26×10^{10} (112)	20,966 (9,510)	50,300	8.26×10^{10} (112)	20,966 (9,510)
Design	63,400	1.84×10^{11} (249)	42,274 (19,175)	60,900	9.00×10^{10} (122)	25,510 (11,571)
Construction	80,879	2.80×10^{11} (380)	72,757 (33,002)	47,100	3.25×10^{10} (44)	7,628 (3,460)
Operation & Maintenance	50,000	1.50×10^{11} (203)	32,628 (14,800)	200,000	7.01×10^{10} (95)	17,066 (7,741)
Total	244,579	7.07×10^{11} (959)	168,625 (76,487)	358,300	2.77×10^{11} (375)	71,240 (32,314)

Example Problem 6.2

Determine the effective horizontal force on a retaining wall 10 ft tall of two types of fill material on unsaturated fill material. Compare blast furnace slag and well-graded gravel, assuming both are coarse-grained soils. State any assumptions that you make.

Since the fill material is unsaturated, Equation 6.8 can be simplified to read:

$$P_o = (P_{A1}) = \left(\frac{1}{2}\sigma_1' K_o H_1\right)$$

H1 is given as 10 ft, so only σ_1' and K_o need to be calculated for each material. For σ_1' the unit weight of each material needs to be provided. Assuming the average value of 95 pcf for the blast furnace slag (Table 6.1) and an average value of 130 pcf for the well-graded gravel (Table 6.2):

$$\text{Slag} \rightarrow \sigma_1' = \gamma \times H_1 = 95 \text{pcf} \times 10 \text{ft} = 950 \text{ psf}$$

$$\text{Gravel} \rightarrow \sigma_1' = \gamma \times H_1 = 130 \text{ pcf} \times 10 \text{ ft} = 1,300 \text{ psf}$$

Next, the earth pressure coefficient at rest is used (again using average internal friction angles from Tables 6.1 and 6.2 for each material, acknowledging that these two materials are both coarse-grained soils:

$$\text{Slag} \rightarrow K_o = 1 - \sin\varphi' = 1 - \sin 42.5 = 0.324$$

$$\text{Gravel} \rightarrow K_o = 1 - \sin\varphi' = 1 - \sin 70 = 0.060$$

Finally, we can solve for the two effective horizontal forces on the wall:

$$\text{Slag} \rightarrow P_o = (P_{A1}) = \left(\frac{1}{2}\sigma_1' K_o H_1\right) = \left(\frac{1}{2} \times 950 \text{psf} \times 0.324 \times 10 \text{ft}\right) = 1,540.9 \text{lb/ft}$$

$$\text{Gravel} \rightarrow P_o = (P_{A1}) = \left(\frac{1}{2}\sigma_1' K_o H_1\right) = \left(\frac{1}{2} \times 130 \text{psf} \times 0.0.060 \times 10 \text{ft}\right) = 392.0 \text{ lb/ft}$$

6.3 MECHANICAL STABILIZED EARTH WALLS

As more retaining walls were designed and installed, a new concept was developed that used the mass of the soil behind the wall to maintain the shape of the soil mass. This concept is referred to as a mechanical stabilized earth (MSE) wall. MSE walls are built in layers, where 1–2 foot lifts of select backfill are placed and compacted, and strips are laid on the soil. These strips can be made out of metal, geogrid, or geotextile. A geogrid is generally made out of a rigid plastic (i.e. polyester) that is arranged in a grid pattern which allows for soil or aggregate interlock within the grid while still increasing the tensile strength of the soil or aggregate. A geotextile is more of a fabric that not only provides tensile strength from friction between the soil and fabric, but can also filter and drain water, or separate different materials without fear of contamination. Both geogrids and geotextiles fall within the geosynthetic family of materials.

After the geosynthetic is laid, a second lift of soil is placed, and a second layer of strips is placed. As the wall moves upward, the soil either slowly steps back with stone facing or is vertical, with typically a precast concrete panel facing. The concrete panels are pinned to the strips, which means the panels are hanging from the strips which are embedded in the earth. The weight of the soil provides friction for the strips and prevents the soil wall from failing. More details regarding standard practice for construction of MSE walls, specifical procedures, and components to be used for soil reinforcement, and connections to panels, as well as selection and placement of granular backfill, can be found in ASTM A1115. Figure 6.8 shows various schematics of MSE walls, Figure 6.9 is a picture of a very steeply stepped MSE wall in Northwest Arkansas, and Figure 6.10 shows an MSE wall with concrete panels being constructed in Little Rock, AR, with a close-up of the back of a concrete panel with ties for the metal strips (a) and an overview with select fill (b). Note in Figure 6.8b, there is a narrow continuous footing underneath the concrete panels, which only provides support for the panels.

On both stepped and vertical MSE walls, the stone facing and concrete panels are mainly for aesthetic purposes and for preventing erosion of the retained soil. Also of interest, in Figure 6.10b, is the sheet piles that are behind the MSE wall being constructed. The pictures of Figure 6.10are of I-630 in Little Rock, AR. The MSE wall was being constructed in order to keep the entire width of the interstate at the same grade, the westbound lanes needed to be elevated significantly. The sheet piles just behind the MSE wall are holding up the existing soil material, while the MSE wall is being constructed. At the end of the construction, the sheet piles will be removed and the MSE wall will hold the exiting soil material in-place. This demonstrates the importance of recognizing temporary structures on construction sites that are necessary to maintain functionality of the infrastructure during the construction phase.

Another important discussion point when considering MSE walls is their social benefit, especially when considering vertical MSE walls. Many vertical MSE walls are used in locations where there is little excess space on either side of the project. On roadways,

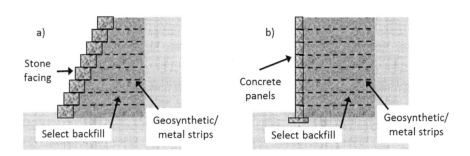

FIGURE 6.8   Stepped (a) versus vertical (b) MSE walls (credit: A Braham)

FIGURE 6.9   Stepped MSE wall (credit: A Braham)

this is termed the right-of-way (ROW). State agencies must design and construct their infrastructure within the ROW, or they need to purchase land adjacent to the ROW. Many existing highways within urban centers are over capacity with very poor Level of Service. However, due to economic and social issues, two of the three pillars of sustainability, it is not feasible to acquire land adjacent to the ROW. Therefore, vertical MSE

(a)   (b)

FIGURE 6.10   a) MSE wall construction concrete panel with ties for the metal strips (credit: A Braham). b) MSE wall construction overview with select fill (credit: A Braham)

walls can fully utilize ROW space, even when there are elevation changes along the project. In Figure 6.10b, a vertical MSE wall was probably constructed because AHTD could not acquire additional land as I-630 passed through Little Rock.

MSE walls must satisfy both internal and external stability. Internal stability governs reinforcement spacing, while external stability governs reinforcement length. Internally, the reinforced soil structure (the select backfill plus strips) must be coherent and self-supporting under its own weight and under any externally applied forces. The reinforcements should not fail either in tension or by pulling out of the select backfill. Externally, the structure must resist overturning (or toppling), sliding at the base, sliding below the base (deep-seated failure), or global instability (bearing capacity failure. In general, the length of reinforcement is 70–80% of the wall height, and the wall is embedded approximately 5–10% of the wall height.

While there are entire books written about the design of retaining walls (Brooks and Nielsen, 2013), it is worthwhile to have a single, simple example of vertical MSE wall design in order to get the flavor of more advanced design principles. For example (Abramson et al., 1996), when designing an MSE wall that utilizes galvanized steel strip reinforcement, the strip length must be long enough in order to support the concrete panel wall and provide a stable mass. The minimal length (L_{min}) of the steel strip can be calculated using Equation 6.11.

$$L_{min} = \frac{F_s KS\Delta H}{2W \tan \delta} \qquad (6.11)$$

where
F_s =safety factor (usually 1.5–2.0)
K   =earth pressure coefficient
S   =horizontal spacing of steel strips
ΔH =vertical spacing of steel strips
W   =width of steel strips
δ =angle of friction of backfill (note, in MSE wall design δ is used instead of the traditional φ that is usually used for the angle of friction in soil applications).

According to Abramson et al., typical dimensions for the horizontal spacing of steel strips (S) is 24 inches, the vertical spacing of steel strips (ΔH) is 10–12 inches, and the width of the steel strips (W) is three inches.

Example Problem 6.3

Using the same two materials and conditions as in Example Problem 6.3, determine the minimum strap length for an MSE wall. State any assumptions you must make.

Utilizing Equation 6.11, the factor of safety will be assumed to be in the middle of the given range, so F_s = 1.75. For slag, K = 0.324 and for gravel, K = 0.060.

Finally, assuming the strap spacing and width that Abramson utilized, S = 24 inches, ΔH = 11 inches (the average value), and W = 3 inches. Therefore, the minimum strap lengths can be estimated:

$$Slag \rightarrow L_{min} = \frac{F_s KS\Delta H}{2W \tan \delta} = \frac{1.75 \times 0.324 \times 24 \, in \times 11 \, in}{2 \times 3 \, in \times \tan 42.5} = 27 \, in$$

$$Gravel \rightarrow L_{min} = \frac{F_s KS\Delta H}{2W \tan \delta} = \frac{1.75 \times 0.060 \times 24 \, in \times 11 \, in}{2 \times 3 \, in \times \tan 70} = 1.7 \, in$$

This calculation shows how for a relatively short wall, and with an especially cohesive fill material (gravel), the use of metal strips may not be most appropriate.

6.4   GEOTHERMAL ENERGY FOUNDATIONS

One of the United Nations' 17 goals for sustainable development is to ensure access to affordable, reliable, sustainable, and modern energy. One of the focuses on achieving this goal includes pursuing the increased use of renewable energy. In 2015, 17.5% of final energy consumption came from renewable energy sources (United Nations, 2018). While this is progress, there is significant room for development in terms of achieving the goal of increased use of renewable energy.

Geothermal energy is a non-polluting, renewable energy source which reduces the fossil fuel energy demand, thus reducing CO_2 emissions (Fragaszy et al., 2011). Due to stability of temperature when compared to outside air temperatures, the ground below a depth of about 20 ft (6 meters) can be utilized for heating and cooling buildings (Olgun et al., 2012). In the long term, these systems have been shown to result in energy cost savings of up to 80% in typical buildings (Hamada et al., 2001). Additionally, geothermal systems can be operated without risk as the low temperature and pressure in the heat carrier circuits are considered to be more hygienic and quieter than conventional air conditioning systems, and reduce dependence on external energy imports which can be dependent on economic or political situations (Brandl, 2016).

Previously, geothermal boreholes were utilized to heat and cool spaces. Although recently, these systems have expanded to include the use of building foundation elements for heat exchange. The result is a hybrid system with geothermal loops integrated into deep foundation elements and connected to a geothermal heat pump. These deep foundation elements can include energy piles, energy diaphragm walls, and energy tunnels.

6.4.1   GEOTHERMAL ENERGY PILES

Energy piles, which were first developed in Austria in the 1980s, are deep foundation elements designed to utilize the consistent temperature of the ground in order to efficiently heat and cool buildings (Olgun et al., 2012). Pile foundations are used to transfer loads into deep layers of firm soil which can sustain loads where soil at or near the surface has inadequate bearing capacity or settlement issues may occur. Therefore, because the ground temperature remains relatively constant (50–75°F or 10–24°C) after a depth of about 20 ft (6 meters) in most regions of the United States, piles which are already in place at these depths and greater to support a structure can also be used as heating and cooling elements (Olgun *et al.*, 2012). As a result, initial costs associated with drilling boreholes to install geothermal energy systems are reduced by utilizing planned structural elements for the dual purpose of heat exchange and foundation support (Abuel-Naga *et al.*, 2014).

Geothermal energy is harnessed by circulation tubes within the piles and a geothermal heat pump acting as heat exchangers. Heat energy is circulated through these tubes with water or antifreeze and is either fed into the ground for cooling in the summer or withdrawn from the ground for heating during winter. This tubing generally consists of U-tube pipes which are fitted into the steel reinforcement cage within the pile, as shown in Figure 6.11. Due to the good thermal conductivity and thermal storage capacity of concrete, piles act as an ideal medium for heat absorption in the ground (Brandl, 2006). Although, Abdelaziz *et al.* found the thermal conductivity of the in-situ soil actually has a more significant effect on the heat exchange capacity of the energy pile, so this should be considered during design (2011). A geothermal

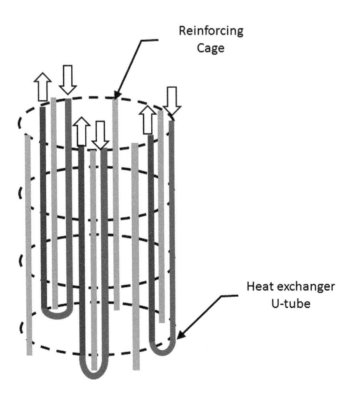

FIGURE 6.11   Schematic of circulation tubes within energy pile (credit: S. Casillas)

heat pump, similar to heat pumps used in residential and commercial applications, then performs the heat exchange.

In terms of design, there is no robust, standardized design method for energy piles despite the increased usage in Europe. Design has been based on empirical considerations in many cases, and factors of safety typically used for piles are increased considerably (Abuel-Naga *et al.*, 2014). As a result, the possible over design of energy piles may lead to unnecessary increases in cost during construction. Researchers, such as Knellwolf *et al.*, have proposed design methods which provide a starting point for future research but rely heavily on significant assumptions and simplifications (2011). The Ground Source Heat Pump Association in the United Kingdom developed standards for energy pile design, installation, and materials (GSHPA, 2012). Unfortunately, these procedures are based on results obtained for one type of soil and therefore may not be appropriate for different soil conditions (Abuel-Naga *et al.*, 2014). Overall, more research is needed to develop a comprehensive design method for energy piles to maximize performance, energy efficiency, and cost-efficiency.

While energy piles have been successfully implemented in some instances, concern about the potential impact of temperature cycles on the load capacity and settlement issues associated with a pile has led to reluctance in accepting this technique (Abuel-Naga *et al.*, 2014). Heating and cooling cycles would lead to expansion and contraction of both the energy pile and surrounding soil, affecting the pile–soil interaction. This creates the potential for unwanted consequences, such as additional settlement of the structure supported by the pile, tensile axial stresses, large compressive axial stresses, or mobilization of a limited resistance on the pile shaft.

Therefore, in addition to considering the structural function of the pile under cyclic temperature conditions, it is imperative to design an energy pile that incorporates a thorough understanding of the thermo-mechanical properties of the soil in which the pile is placed as well (Abuel-Naga *et al.*, 2014). Previous research has sought to better understand the performance of soils in response to changes in temperature as it may pertain to pile–soil interactions. For instance, the soil plasticity index (PI) is believed to indicate the magnitude of thermally induced volumetric strain of normally consolidated clays as these strains have been attributed to physiochemical interactions between clay particles. Therefore, as PI increases, the magnitude of thermally induced volumetric strain also increases as the swell potential would. For example, a clay soil with a PI of 18 would be expected to be more susceptible to thermally induced volumetric strain than a clay soil with a PI of 7. In terms of shear performance, Abuel-Naga *et al.* found that undrained shear strength of normally consolidated clay increases as temperature increases or after the soil has been subjected to temperature cycling (2006, 2007). Many researchers have also generally reported hydraulic conductivity of soil increases as temperature increases (Burghignoli *et al.*, 2000; Delage *et al.*, 2000; Houston and Lin, 1987; Morin and Silva, 1984; Towhata *et al.*, 1993). Finally, as an undrained soil is heated, pore water pressure has been shown to increase, which can induce additional stresses in the soil, even to the point of failure if the stress state reaches the shear strength of the soil (Hueckel and Pellegrini, 1992). Each of these soil properties and behaviors could influence the ways in which the pile interacts with soil over time and during temperature cycles.

6.4.2 Geothermal Diaphragm Walls

Diaphragm walls are typically used to support deep excavations when alternative techniques, such as piles, are not appropriate. Common applications include deep basement walls, subway tunnels, and areas with high water tables (Brandl, 2016). Like energy piles, closed pipes placed within the panels of the diaphragm wall serve to circulate heat carrier fluid through the structure, transferring heat from the building to the ground or vice versa. These pipes then connect into a heat pump which circulates heat into the above structure, as shown in Figure 6.12. According

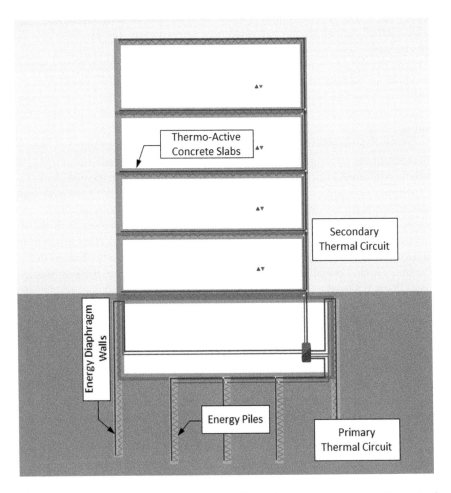

FIGURE 6.12 Cross section of a building with geothermal diaphragm wall (credit: S. Casillas)

to multiple case studies, geothermal diaphragm walls are typically 2.6 ft to 3.9 m (0.8–1.2 meters) wide and 32.8 ft to 131.2 feet (10–40 meters) deep (Gaba *et al.*, 2003; Burland *et al.*, 2012).

The energy efficiency of geothermal diaphragm walls is influenced by a number of factors which are not considerations for energy piles. Both the length of the wall as well as the area of the wall embedded in the soil and above ground significantly influence energy efficiency (Di Donna *et al.*, 2017). The portion of wall which makes contact with soil on one side will experience different rates of heat exchange than the portion of wall exposed to the open air. The energy efficiency of a geothermal diaphragm wall will also be influenced by the configuration of heat transfer pipes within the wall. This includes considerations like length of concrete cover between the heat transfer pipes and edge of wall, spacing of the pipes, and whether pipes are installed on both sides of the wall or only one. In Austria, heat transfer pipes are typically installed on both sides of the wall – the interior and the exterior which makes contact with the soil. However, pipes are only placed in the exterior side for the portion of wall which is embedded in soil (Di Donna *et al.*, 2017). Heat transfer pipes are fixed to the wall's steel reinforcement cages prior to construction whenever possible. As discussed for energy piles, the thermal conductivity and thermo-mechanical behavior of the soil in which the diaphragm wall is constructed also influence energy efficiency.

When extending the Metro line U2 in Vienna, Austria, geothermal diaphragm walls were installed in stations for heating and cooling purposes. This was the first full-scale application of the use of geothermal energy structure technology in metro engineering (Brandl, 2016). Prior to construction, feasibility studies using numerical analyses of temperature flow in the ground concluded the extraction and storage of geothermal energy would have very limited influence on the soil and groundwater surrounding the walls, indicating unfavorable thermal effects such as the settlement issues and addition of stresses in the soil would not be a concern. The soil profile in which one station was constructed was comprised of layers of sandy silt, well-graded gravel, sandy gravel, clayey sand, and silty sand, from the surface downward, respectively. This station featured a 20,075 ft^2 (1865 m^2) geothermal diaphragm wall. According to initial observations and measurements, outfitting these metro stations and metro lines with geothermal foundation structures has proven to be both environmentally beneficial through the harvesting of geothermal energy and economically viable (Brandl, 2016).

6.4.3 ENERGY TUNNELS

The third type of geothermal foundation element will be discussed is an energy tunnel. Compared to piles and diaphragm walls, tunnels have a much larger surface area and therefore larger volume of soil with which heat can be exchanged. Heat exchanged through an energy tunnel can be moved to the surface using pipes placed in ventilation shafts of the tunnel. Another form of heat exchange for energy tunnels is to cool the tunnels.

Depending on the use of a tunnel and local conditions, energy tunnels can be classified as cold or hot. Cold tunnels are built in locations where air temperature is low (about 59°F/15°C) all year, and trains pass through too infrequently to increase the temperature within the tunnel. These tunnels do not have a large effect on the temperature of the soil through which the tunnel runs. Hot tunnels, then, have high internal temperatures, around 86°F (30°C). Hot tunnels are typically urban tunnels with underground railways or subway systems, featuring rapid frequency of trains. The braking and starting of these trains adds to the heat input from the tunnel into the surrounding soil. Available geothermal energy can be exploited in the case of both cold and hot tunnels (Barla and Di Donna, 2018).

Based upon construction methods and materials, tunnel linings can be optimized to maximize the geothermal energy transfer. Initially, heat transfer pipes were attached to non-woven geosynthetics offsite prior to tunnel construction. When mechanized tunneling began, however, tunnel lining began to be fabricated as pre-cast segments which were installed by the tunnel boring machine (Barla and Di Donna, 2018). This advancement introduced the opportunity to optimize the lining segments for heat transfer by placing hydraulic circuits within the concrete segments. The pipes within each segment can be connected through hydraulic connections, forming lining ring circuits. These circuits then connect into a main conduit pipe which carries the heat exchange fluid from the tunnel lining to the heat pump if moving heat away from the tunnel or vice versa if brining heat into the tunnel (Barla and Di Donna, 2018). Optimization can be accomplished through the specific configuration and spacing of these pipes within the tunnel lining segments.

When designing an energy tunnel, two additional considerations arise over typical tunnel design criteria. First, available heat energy should be quantified to understand the energy efficiency associated with this tunnel. This will help designers identify the viability of constructing an energy tunnel for a given project. Quantifying efficiency also aids in the design of heat carrier piping throughout the tunnel lining. The second consideration is ensuring the geothermal energy capture elements do not induce negative long-term effects on the structural integrity of the tunnel. While there are potential environmental and economic savings associated with exploiting the available geothermal energy with tunnels, ensuring the safety of individuals using these tunnels is most important for any designers working on these projects. Therefore, a thermo-mechanical coupled analysis is needed to understand the thermal efficiency and structural integrity of the tunnel (Barla and Di Donna, 2018). As with piles and diaphragm walls, this evaluation not only considers the tunnel structure itself but also the response of the surrounding soil to the thermo-mechanical loading which could be induced.

Example Problem 6.4

Designers in Fayetteville, Arkansas, are considering converting some of the piles which will serve as foundations on upcoming projects to energy piles. When evaluating this design choice, a geotechnical engineer was sent to two different building sites being considered to obtain a soil profile. The geotechnical report showed that the bulk of the soil layers in which the pile would be placed was classified as a fat clay for site number one and a lean clay for site number two. The fat clay was reported as having a liquid limit (LL) of 60, and the lean clay had an LL equal to 40. Using ASTM D2487-11, determine the plasticity index (PI) value for each of these soils assuming both plot on the "A" line. Based upon the PI values you report, make a recommendation as to which site may experience fewer negative consequences associated with the thermal cycles expected for an energy pile.

- First, determine the soil classification for each of these soil types using the Unified Soil Classification System (USCS) chart found in ASTM D2487-11. A fat clay is given the symbol CH. Lean clays are CL.
- Second, refer to the chart which plots PI versus LL in ASTM D2487-11, known as the PI chart. Referencing the USCS and regions shown in the PI chart, identify regions in which CH and CL soils typically fall.
- Now using the LL values given, determine the PI associated with each of these soils. For the CH on site one with an LL = 60, PI would be 30. For the CL on site two with an LL = 40, PI is approximately 15.
- Comparing the PI values of the two soils, the PI of the CH on site one is double that of the CL on site two. Therefore, it is to be expected that the CH would be more susceptible to volumetric strain induced by thermal cycles. This could cause settlement issues for the building supported by the energy piles. So, based on PI alone and the potential for thermally induced volumetric strain, site two with a CL soil would be a better fit for implementing energy piles within the building foundation.

HOMEWORK PROBLEMS

1. Using the data from Example Problem 6.1, calculate the void ratio, the saturated unit weight, and the porosity.
2. Create a phase diagram for reclaimed concrete. Assume a specific gravity of 2.678, a soil volume of 1 ft³ and a total volume of 1.3 ft³. In addition, calculate the void ratio, the saturated unit weight, and the porosity. State any assumptions that need to be made.
3. Determine the effective horizontal force on a retaining wall 20 ft tall of two types of fill material on unsaturated fill material for the top 10 ft and saturated fill material for the bottom 10 ft. Compare blast furnace slag and well-graded gravel, assuming both are coarse-grained soils. State any assumptions that you make.
4. Bioengineered slopes seem to have many beneficial characteristics. However, there are challenges during the design, construction, and maintenance of bioengineered slopes. Identify what you believe the two largest challenges are, and discuss using the format provided under the sidebar in Chapter 1 "Writing a High-Quality Essay."
5. Determine the minimum strap length for an MSE wall with lean clay as the fill material (Note: assume you can estimate the earth pressure coefficient using the equation given for coarse-grained soils). The wall will need to be 20 ft tall. State any assumptions you must make.
6. In total, six types of retaining walls were explored. List the six types, and choose which type you think is most sustainable. Give one example of

economic, environmental, and social reasons for your choice. Discuss using the format provided under the sidebar in Chapter 1 "Writing a High-Quality Essay."

7. A number of considerations and factors which can influence the success of installing a geothermal structure such as an energy pile were discussed in this chapter. Using the format provided under the sidebar in Chapter 1 "Writing a High-Quality Essay," describe at least three considerations an engineer should evaluate prior to electing to use a

8. The design team for a public library being built in Dallas is considering utilizing energy piles rather than traditional piles as foundation elements. However, areas surrounding Dallas are known to have highly expansive clay. Therefore, the design team has decided that energy piles will only be used if the plasticity index (PI) of the soil on site is less than 25. The geotechnical engineer obtained a soil profile for the project site. The report indicated the bulk of the soil layers in which the pile would be placed are poorly graded gravel and lean clay. The lean clay was determined to have a liquid limit equal to 30. Using ASTM D2487-11, determine the potential range of plasticity index (PI) values for the lean clay. Based upon the PI values you report and the upper limit selected by the design team, make a recommendation as to whether or not energy piles should be used.

REFERENCES

Abdelaziz, S.L., Olgun, C.G., Martin, J.R. Design and Operational Considerations of Geothermal Energy Piles. In *Proceedings of Geo-Frontiers 2011: Advances in Geotechnical Engineering* (Han J and Alzamora DE (eds)). ASCE Geotechnical Special Publication No. 211, pp. 450–459, Dallas, TX, 2011.

Abramson, L., Lee, T., Sharma, S., Boyce, G. *Slope Stability and Stabilization Methods*. John Wiley & Sons, New York City, NY, 1996.

Abuel-Naga, H.M., Bergado, D.T., Chaiprakaikeow S. Innovative Thermal Technique for Enhancing the Performance of Prefabricated Vertical Drain System. *Geotextiles and Geomembranes*, 2006, 24(6): 359–370.

Abuel-Naga, H.M., Bergado, D.T., Bouazza, A. Thermally Induced Volume Change and Excess Pore Water Pressure of Soft Bangkok Clay. *Engineering Geology,* 2007, 89(1–2): 144–154.

Abuel-Naga, H., Raouf, M. I. N., Raouf, A. M. I., Nasser, A. G. Energy Piles: Current State of Knowledge and Design Challenges. *Environmental Geotechnics*, 2014, 2(4) 195–210.

Barla, M., Di Donna, A. Energy Tunnels: Concept and Design Aspects. *Underground Space*, 2018, 3, 268–276.

Brandl, H. Energy Foundations and Other Thermo-Active Ground Structures. *Géotechnique,* 2006, 56(2): 81–122.

Brandl, H. Geothermal Geotechnics for Urban Undergrounds. *Procedia Engineering*, 2016, 165, 747–764.

Brooks, H., Nielsen, J. *Basics of Retaining Wall Design, A Design Guide for Earth Retaining Structures*. 10th Edition, HBA Publications, 2013.

Burghignoli, A., Desideri, A., Miliziano, S. A Laboratory Study on the Thermomechanical Behaviour of Clayey Soils. *Canadian Geotechnical Journal*, 2000, 37, 764–780.

Burland, J. et al. *ICE Manual of Geotechnical Engineering*. L. Thomas Telford, ed., ICE Publishing, London, 2012.

Chesner, W., Collins, R., MacKay, M. User Guidelines for Waste and By-Product Materials in Pavement Construction. FHWA-RD-97-148, Report Number 480017, Guideline Manual. Federal Highway Administration, 1998.

Delage, P., Sultan, N., Cui, Y.J. On the Thermal Consolidation of Boom Clay. *Canadian Geotechnical Journal*, 2000, 37, 343–354.

Di Donna, A., Cecinato, F., Loveridge, F., Barla, M. Energy Performance of Diaphragm Walls Used as Heat Exchangers. *Proceedings of the Institution of Civil Engineers: Geotechnical Engineering*, 2017, 170, 232–245.

Fragaszy, R.J., et al. Sustainable Development and Energy Geotechnology – Potential Roles for Geotechnical Engineering. *KSCE Journal of Civil Engineering*, 2011, 15(4), 611–621.

Gaba, A.R. et al. Embedded Retaining Walls – Guidance for Economic Design, CIRIA C580, Construction Industry Research and Inf., 2003.

GSHPA. Thermal Pile Design, Installation and Materials Standards. *Ground Source Heat Pump Association*, 2012. www.gshp.org.uk/GSHPA_Thermal_Pile_Standard.html.

Hamada, Y., Nakamura, M., Ochifuji, K., Nagano, K., Yokoyama, S. Field Performance of a Japanese Low Energy Home Relying on Renewable Energy. *Energy and Buildings*, 2001, 33(8), 805–814.

Houston, S.L., Lin, H.D. A Thermal Consolidation Model for Pelagic Clays. *Marine Geotechnology*, 1987, 7, 79–98.

Hueckel, T., Pellegrini, R. Effective Stress and Water Pressure in Saturated Clays During Heating-Cooling Cycles. *Canadian Geotechnical Journal*, 1992, 29, 1095–1102.

Knellwolf, C., Peron, H., Laloui, L. Geotechnical Analysis of Heat Exchanger Piles. *Journal of Geotechnical and Geoenvironmental Engineering*, 2011, 137(10), 890–902.

Morin, R., Silva, A.J. The Effect of High Pressure and High Temperature On Some Physical Properties of Ocean Sediments. *Journal of Geophysical Research*, 1984, 89(B1), 511–526.

Olgun, C. G., Abdelaziz, S. L., Martin, J. R. Long-Term Performance and Sustainable Operation of Energy Piles. *ICSDEC 2012: Developing the Frontier of Sustainable Design, Engineering Construction*, 2012, 534–542.

Rankine, W. On Stability on Loose Earth. *Philosophic Transactions of Royal Society, London, Part I*, 1857, 9–27.

Storesund, R., Massey, J., Kim, Y. *Life Cycle Impacts for Concrete Retaining Walls vs. Bioengineered Slopes, GeoCongress 2008: Geosustainability and Geohazard Mitigation*, New Orleans, LA, March 9–12, 2008, 875–882.

Towhata, I., Kuntiwattanakul, P., Seko, I., Ohishi, K. Volume Change of Clays Induced by Heating as Observed in Consolidation Tests. *Soils & Foundations*, 1993, 334, 170–183.

United Nations. Ensure Access to Affordable, Reliable, Sustainable and Modern Energy. *Sustainable Development Goals*, 2018. www.un.org/sustainabledevelopment/energy/.

7 Application
Structural Sustainability

Don't let schooling interfere with your education.

Mark Twain

Many young aspiring civil engineers enter the discipline because of structures. Whether soaring skyscrapers or elegant bridges, structures is the extroverted side of civil engineering. The area of structures also has many different perspectives for sustainability. Whether looking at materials, design, or evaluation, there are many various applications to incorporate sustainable practices. This chapter will cover two material-related themes, one design theme, and finish with a discussion about how to evaluate the sustainability of structural systems. Specifically, this chapter will cover these four areas:

1. Fly ash
2. Bamboo
3. Design for Adaptability and Deconstruction (DfAD)
4. Cross Laminated Timber

7.1 FLY ASH

It is estimated that there were over 600 coal power plants in the United States that produced over 52 million short tons of fly ash in 2012 (ICF International, 2016). Fly ash is both inorganic and noncombustible, and is a residue of coal after burning in power plants. Fly ash is a pozzolanic material, and can be used as a supplement of Portland cement in Portland Cement Concrete (PCC). During combustion of coal, the volatile matter and carbon are burned off, while the mineral impurities melt and are fused together. These mineral impurities include clay, feldspar, and quartz. The fused material is moved to low temperature zones where it solidifies into spherical particles of glass. The material that falls is bottom ash (from agglomeration), while the rest of the material is light enough to be lifted out with the flue gas stream. This light material is the fly ash, and it is removed from the gas by cyclone separation, electrostatic precipitation, and bag-house filtration.

There are two categories of fly ash: Class C (high calcium) and Class F (low calcium). The mean size of both Class C and F fly ashes is 10–15μm, the surface area is 1–2 m²/g, and the specific gravity is 2.2–2.4. Table 7.1 shows some differences in mineral composition between the two classes. Class C fly ash comes from anthracite and bituminous coal, whereas Class F fly ash comes from lignite and subbituminous

TABLE 7.1
Common Characteristics of Fly Ash

	mg Ca(OH)$_2$ consumed/gram	SiO$_2$ (%)	Al$_2$O$_3$ (%)	Fe$_2$O$_3$ (%)	CaO (%)	Carbon (%)
Class C	500	>30	15–25	<10	20–30	<1
Class F	850	>50	20–30	<20	<5	<5

coal. In general, if the amount of carbon in fly ash is greater than 5%, it is not desirable for use in PCC.

Fly ash with higher amounts of calcium display cementitious behavior, and will react with water to perform hydrates when calcium hydroxide is not available. This reaction benefits PCC as it increases the cementitious binder phase (calcium-silicate hydrates or C-S-H), improving the long-term strength and decreasing the permeability. Fly ash is generally used as a partial substitute for Portland cement.

Not only are there physical differences between the two materials, as seen in Figure 7.1, but the two materials also influence both the fresh and hardened concrete properties. These influences are summarized nicely in a publication put out by the Portland Cement Association (Thomas, 2007), and summarized as follows in the following paragraphs.

When examining fresh concrete, fly ash can affect the workability, water demand, setting time, heat of hydration, and finishing and curing. In terms of workability, the addition of fly ash increases workability, making the concrete easier to place, consolidate, and finish. The increase of workability comes from the fact that fly ash has a relatively high fineness and low carbon content which reduces the need for water. Therefore, in general, less water is needed when substituting either Class F or Class

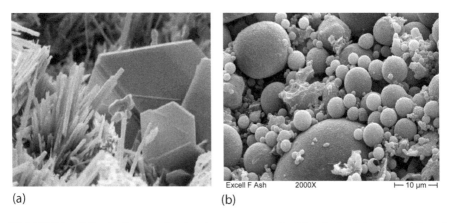

(a)   (b)

FIGURE 7.1   a) Portland cement hardened paste, plates of calcium hydroxide, and needles of ettringite, micron scale (credit: US DOT). b) Fly ash (credit: PMET Lab Service)

C fly ash for Portland cement. A rule of thumb is for each 10% of fly ash substituted for Portland cement, the water can be reduced by 3%.

In addition to the workability and water content, the setting time can also be decreased. In general, low-calcium fly ashes extend both the initial and final set times of PCC. This influence, however, is dependent on ambient temperature. During hotter weather, the extension of time is reduced and may actually become a benefit, whereas in colder weather, the extended set time can cause delays and thus complicate placement and finishing operations. Conversely, higher-calcium fly ashes do not retard setting time as much as lower-calcium fly ashes because of the increase of hydraulic reactivity. However, this trend is more difficult to predict, so a full laboratory evaluation is recommended before incorporating a new fly ash source.

The next fresh PCC characteristic, heat of hydration, is an incentive for using fly ash, especially in mass concrete construction. Large-scale structures, such as dams or large bridge columns, can cure improperly due to the high heat within the concrete mass, which could lead to cracking and other temperature-related damage. However, if early age strengths are not necessary, the use of fly ash can reduce the heat of hydration, thus reducing potentially harmful high temperatures. Studies have shown that replacing approximately half of Portland cement with Class F fly ash, for example, can reduce the maximum temperature in large concrete blocks by almost 30% (Langley *et al.*, 1992).

While the workability, setting, and heat of hydration can all be influenced by fly ash, the final finishing and curing should also be considered. The rate of pozzolanic reaction is slower than the rate of cement hydration, and more care must be taken in order to ensure proper curing. When using fly ash, the concrete should be moist cured for a minimum of 7 days, and ideally, a curing member should be added after 7 days and curing should be extended to 14 days. Table 7.2 summarizes the trends of fresh properties of concrete using fly ash.

In addition to fresh concrete, hardened concrete properties must also be accounted for. Important hardened properties include compressive strength development, permeability, and alkali-silica reaction. In general, the initial compressive strength of concrete is lower with fly ash, and the strength continues to decrease with an increase of fly ash. However, long-term strength is actually increased with the use of fly ash.

The permeability of PCC is important for durability and long-term performance in the field, especially in the presence of chlorides. Chlorides are especially destructive

TABLE 7.2
Fly Ash Impact on Fresh Properties of Concrete

Fly ash type	Workability	Water demand	Setting time	Heat of hydration	Finishing and curing
Class F	Increases	Decreases	Increases	Decreases	Increase
Class C	Increases	Decreases	May increase or decrease	May increase or decrease	May increase or decrease

TABLE 7.3

Fly Ash Impact on Hardened Properties of Concrete

Fly ash type	Compressive strength – short term	Compressive strength – long term	Permeability	Alkali-silica reaction
Class F	Decreases	Increases	Decreases	Decreases
Class C	May increase or decrease	Increases	Decreases	Decreases

to reinforcing steel, and will destroy the passive oxide fill on steel if able to permeate into the PCC. Therefore, it is beneficial to have low permeability. With the addition of fly ash, the permeability decreases in PCC.

Class F fly ash can control damaging alkali-silica reaction (ASR) at intermediate levels of replacement (20–30%). AST occurs when the alkalis in the cement paste react with certain types of silica in the aggregate. The concrete expands during this reaction which causes cracking. While Class C fly ashes are less effective, both classes of fly ash essentially reduce the concentration of alkali hydroxides in the pore solution when fly ash is present. Table 7.3 summarizes the trends of hardened properties of concrete using fly ash.

Overall, fly ash influences both the fresh and hardened properties of PCC, and if properly managed, it can increase the performance of PCC. In addition to these material benefits, fly ash is also very beneficial from both an economic and environmental standpoint.

From an economic standpoint, fly ash has shown to have an advantage over Portland cement. Lippiatt and Ahmad (2004) performed a life cycle cost analysis that incorporated costs from the product purchase onward, which encompasses all out-of-pocket costs. They calculated that both first and future costs would be approximately 10% lower with the incorporation of 35% fly ash into a PCC mix (with 65% Portland cement) compared to a PCC mix containing 100% Portland cement. A second study by Santero *et al.* (2011) anticipated an average savings of approximately $15,000/lane-km when going from 10% to 30% fly ash replacement for Portland cement in PCC pavements. Finally, a study by Lu (2007) found an optimal value of 23% fly ash replacement for use in footbaths and bicycle lanes. These studies all show that replacing virgin Portland cement can potentially save money over the life of the application.

Similarly, environmental benefits of utilizing fly ash have been found. Ondova and Estokova (2014) performed a Life Cycle Assessment on 15% replacement of fly ash for Portland cement, and examined the extraction, production, application, and disposal/recycling phase of PCC. Overall, the research found that utilizing 15% fly ash reduced the Global Warming Potential (GWP) from 1763 to 1668 kg CO_2 equivalent/kg (a reduction of over 5%) and the acidification potential from 3.4 to 3.2

kg SO_2 equivalent/kg (a reduction of over 6%). A second study used the PaLATE LCA tool, which incorporates production of materials, construction, maintenance, and end-of-life processes (Ahlman *et al.*, 2015). The research took findings from six state Department of Transportations (Wisconsin, Minnesota, Illinois, Pennsylvania, Colorado, and Georgia), and found that the environmental benefits of the use of fly ash provided 81% savings in energy, 88% savings in water consumption, and 82% in CO_2. Overall, these studies show significant environmental benefits of fly ash as a replacement of Portland cement.

Example Problem 7.1

A local contractor is investigating replacing 30% virgin Portland cement with fly ash. The concrete with 100% concrete has a compressive strength of 33 MPa at three days and a compressive strength of 52 MPa at 90 days. The concrete with 30% fly ash has a compressive strength of 28 MPa at 3 days and 60 mPa at 90 days. Using the common relationship between compressive strength and modulus of elasticity, calculate the secant modulus of elasticity (E_c) at these four strength levels.

The common relationship is: $E_c = 0.043 w_c^{1.5} \sqrt{f_c'}$, where the unit weight of concrete is usually assumed to be 2320 kg/m$_3$, giving: $E_c = 4730\sqrt{f_c'}$. Therefore:

100% Portland cement, 3 day cure $\rightarrow E_c = 4730\sqrt{33\ \text{MPa}} = \underline{27.2\ \text{GPa}}$

100% Portland cement, 90 day cure $\rightarrow E_c = 4730\sqrt{52\ \text{MPa}} = \underline{34.1\ \text{GPa}}$

30% fly ash, 3 day cure $\rightarrow E_c = 4730\sqrt{28\ \text{MPa}} = \underline{25.0\ \text{GPa}}$

30% fly ash, 90 day cure $\rightarrow E_c = 4730\sqrt{60\ \text{MPa}} = \underline{36.6\ \text{GPa}}$

7.2   BAMBOO

Wood is a common building material, especially in the United States and Europe, with the added benefit of being a renewable resource and a location for carbon storage. The United States has approximately 746 million acres of forest land that provide collection of carbon through absorption. With the continued use of carbon dioxide emissions, these types of "terrestrial pools" are beneficial in absorbing carbon from the atmosphere (EPA, 1995). However, another renewable resource that could be used as a building material is bamboo. Bamboo, like wood, is renewable and has mechanical properties similar to timber (Widenoja, 2007). In fact, according to Sharma et al., bamboo has a faster growth rate and shorter harvest cycle versus wood, and has four times the carbon density versus spruce forests. Wood is most frequently found in the northern hemisphere (North America, Europe, Russia), while the majority of developing areas in the world are in the equatorial regions or in the southern hemisphere, locations where bamboo is more common than wood. The general structure of bamboo is similar to wood, as it is an anisotropic material, where the properties vary in the longitudinal, radial, and transverse directions. The structure of bamboo, where longitudinal fibers align with a lignin matrix, is what causes this anisotropic behavior. These longitudinal fibers are divided by solid diaphragms along the longitudinal length, as seen in Figure 7.2.

Overall, there are 1,200 species of bamboo worldwide, with a variation in both geometrics and mechanical properties, making it difficult to design connects and joints suitable for various sections of construction material, such as columns, beams, or other permanent, load-bearing structures. However, bamboo composites are of interest because of the standardization of shape, and low variability of material properties.

FIGURE 7.2   Solid diaphragms dividing the longitudinal fibers (credit: Alain Van den Hende)

Similar to plywood and particle board, bamboo can be broken down and reassembled in a more beneficial form (Sharma *et al.*, 2015). There are two types of bamboo composites available, laminated and scrimber. Laminated bamboo preserves both the longitudinal and culm matrix. The bamboo is split and planed, bleached, and caramelized. After lamination, it is pressed to form a board product. This process uses approximately 30% of the raw inputs, as much of the input material is lost during the planning process.

The second bamboo composite is scrimber, which maintains only the longitudinal matrix. The scrimber bamboo is produced by weaving the longitudinal strands, crushing the woven strands, and saturating in resin. This compressed, dense block maintains the longitudinal direction of the fibers and the resin matrix connects the fiber bundles. A benefit to this technique is that the scrimber process utilizes approximately 80% of the raw inputs.

Sharma *et al.*, performed a study that compared various material properties of both scrimber and laminated bamboo with raw bamboo, spruce lumber, and laminated veneer lumber. This allowed for direct comparison between not only the two different forms of composite bamboo (laminated and scrimber), but also the raw materials alone (bamboo and spruce) and two different laminated materials (lumber and bamboo). The results are shown in Figure 7.3a–7.3d. Note, the compressive, tensile, and shear strength are all measured parallel to the materials' primary orientation.

In Figure 7.3, scrimber bamboo has higher values in all four mechanical properties versus laminated (except the shear strength, which is essentially the same in the two materials). Therefore, it would appear that since scrimber bamboo has better performance and uses a higher percentage of raw inputs, it would be the preferable method of utilizing composite bamboo. When comparing the laminated bamboo to the laminated wood, the

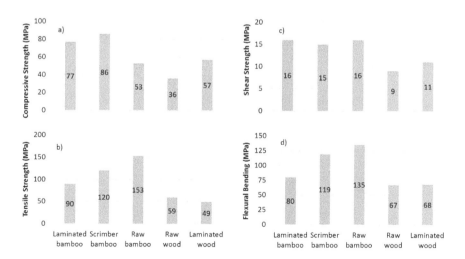

FIGURE 7.3 Mechanical properties of laminated bamboo, scrimber bamboo, raw bamboo, raw wood, and laminated wood (data from Sharma *et al.*, 2015)

laminated bamboo has higher values in all four mechanical properties without exception, indicating that it is a stronger and more versatile material. Finally, raw bamboo also has higher values in all four mechanical properties versus raw wood. From this data set, it would appear that bamboo is a stronger material than wood. However, when thinking about sustainability, it is also critical to think of the cost (economic pillar) and the processing requirements (economic and environmental pillar). In the United States, the costs and processing requirements would be higher for bamboo since the supply is lower, but in southeast Asia, where bamboo is more commonly used than wood, the economic and environmental impacts are most likely lower.

The discussion up to now has focused on comparing bamboo to wood, but there are also applications where bamboo is a reasonable replacement of steel. A common form of scaffolding in the United States and Europe is steel, but in China and Southeast Asia, bamboo is much more prevalent as a scaffolding material. For example, Figure 7.4 shows a construction site in Hefei, Anhui province, China, that utilized bamboo as a scaffolding material.

A mode of failure that is essential when constructing scaffolding is buckling failure. When considering elastic buckling stresses, the calculations for bamboo and steel are a bit different, but they have a similar starting point. Buckling, regardless of material, is generally represented by Euler's formula, which is seen in Equation 7.1:

$$F_e = \frac{\pi^2 E}{\left(KL/r\right)^2} \tag{7.1}$$

FIGURE 7.4   Bamboo used as a scaffolding material in Hefei, China (credit: A Braham)

where:
  F_e = elastic buckling force
  E = modulus of elasticity
  K = column effective length factor
  L = unsupported length of column
  r = radius of gyration

The column-effective length factor is dependent on the end conditions. When both ends are pinned, $K = 1$. When both ends are fixed, $K = 0.50$. If one end is fixed and the other end is pinned, $K = 0.7071$. Finally, if one end is fixed and the other end is free to move laterally, $K = 2.0$.

While Equation 7.1 provides the foundation for both the bamboo and steel buckling equation, when considering bamboo, the alpha value is placed in front of the equation, creating Equation 7.2:

$$F_e = \frac{\alpha \pi^2 E}{(KL/r)^2} \tag{7.2}$$

where α is a function of the second moment of area, and depends on the moisture content. Alpha ranges from 1.00 to 2.35, and increases as the moisture content increases.

Example Problem 7.2

A firm in Central Arkansas would like to explore using bamboo as a substitute for steel for scaffolding. However, they are concerned about the load-carrying ability of the bamboo versus the steel. Assume that both materials are pinned or connected at each end; the length of the column will be 12 ft and the radius of gyration of each material will be 1.25 in. Also, assume that the modulus of elasticity of the steel is 28.8×10^6 psi. Compare the steel to bamboo in a wet state ($\alpha = 2.35$, $E = 0.97 \times 10^6$ psi) and in a dry state ($\alpha = 2.35$, $E = 1.50 \times 10^6$ psi).

Using Equation 7.1, the elastic buckling force can be calculated for steel:

$$\text{Steel} \rightarrow F_e = \frac{\pi^2 E}{(KL/r)^2} = \frac{\pi^2 \times 28.8E^6 \text{ psi}}{(1.0 \times 24 \text{ in}/1.25 \text{ in})^2} = \underline{771.1 \text{ ksi}}$$

Using Equation 7.2, the elastic buckling force can be calculated for the bamboo:

$$\text{Bamboo dry} \rightarrow F_e = \frac{\alpha \pi^2 E}{(KL/r)^2} = \frac{1.00 \times \pi^2 \times 1.50E^6 \text{ psi}}{(1.0 \times 24 \text{in}/1.25 \text{ in})^2} = \underline{40.2 \text{ ksi}}$$

$$\text{Bamboo wet} \rightarrow F_e = \frac{\alpha \pi^2 E}{(KL/r)^2} = \frac{2.35 \times \pi^2 \times 0.97E^6 \text{ psi}}{(1.0 \times 24 \text{ in}/1.25 \text{ in})^2} = \underline{61.0 \text{ ksi}}$$

7.3 DESIGN FOR ADAPTABILITY AND DECONSTRUCTION (DFAD)

As discussed in Chapter 2, recycling, remanufacturing, and reuse are three concepts related to the six life cycle stages of infrastructure. Following this order of concepts, there is an increase of impact for relative sustainability. While architects and mechanical engineers may be thought to play a larger role in sustainable design decisions of a structure, structural engineers can still play a pivotal role in ensuring a structure and its components can be recycled, remanufactured, or reused (Webster, 2007). Design for Adaptability and Deconstruction (DfAD) is a structural design strategy which greatly increases the probability the life of the structure will be extended (adaptability) and the materials will be recycled, remanufactured, or reused at the structure's end of life (deconstruction).

The adaptability portion of DfAD deals primarily with the duration of life, as increasing the versatility of a building can make it suitable for multiple purposes and tenants or owners. Adaptability measures the ease of modifying a building during the use stage of the life cycle (the fourth stage). This allows a structure to accommodate various future uses, requiring minimal cost and effort in order to do so (Webster, 2010). Adaptability can be divided into flexibility, which is the ability to make minor changes to a space which affects the usage, and convertibility, which is the ability to accommodate changes in use (Moffatt and Russell, 2001). According to a study conducted in Minnesota, about 40% of demolitions occurred because the physical condition of the building required it (Athena, 2004a). The remaining 60% of demolitions were completed due to non-structural issues of neighborhood redevelopment and suitability for a designated purpose. This highlights the necessity of designing a structure with adaptability in mind. A building's inability to be flexible and convertible may lead to the costs of adapting exceeding costs of designing and constructing a new building entirely. This likely results in materials comprising the original building becoming waste.

Designing for deconstruction requires the building's end of life (the fifth life cycle stage) be considered during the design phase. Deconstruction is a demolition method whereby a structure is carefully and methodically disassembled in order to salvage the maximum number of components (2007, Webster). This methodology is also referred to as "design for disassembly." Ideally, this approach encourages not only recycling but also material reuse. Recycling materials is beneficial when compared to using more virgin materials; however, recycling can require a large quantity of energy and therefore contribute significantly to pollution. Reuse does not require any sort of re-manufacturing, so the environmental impact is even lower than recycling. For example, if a steel column is removed from one building, refabricated, and then placed into a new building, this is considered reuse. If, on the other hand, the column is sent to a mill and merged with other scrap steel, rolled into a new section, fabricated into a new column, and then installed in a new building, the steel is recycled.

DfAD is considered to be a sustainable design approach due to a number of environmental and economic benefits (Webster, 2007; Webster, 2010). First, considering the environmental pillar, increasing the length of life and functionality of a building decreases the need for new construction. Therefore, the environmental impacts associated with new construction are decreased, such as site disturbance, habitat destruction, toxic emissions, and generation of greenhouse gases (GHG) through the manufacture and transport of new materials. In general, DfAD helps close the material loop by encouraging the reuse of materials by salvaging used materials at deconstruction. The reuse of materials can help significantly reduce energy consumption and GHG emissions. For example, using salvaged brick can reduce the environmental impact as measured by energy consumption and GHG emissions by up to 98% (Athena, 2004b). Finally, because materials are being salvaged for reuse or recycling, less waste is generated which would need to be hauled away and placed in a landfill.

The use of DfAD can also result in benefits as well, directly impacting the economic pillar. Designing a structure which can be easily adapted can simplify maintenance and repair thereby reducing costs. Construction and adaptation costs are reduced through the use of standard connections and repeating geometries. Using salvaged and recycled materials can decrease material costs and energy costs which may be associated with the manufacture of new materials. Overall, the value of the building may be increased due to materials having a higher salvage value at the end of life due to easy extraction and reuse. Buildings can receive LEED Green Building Rating system credits for the use of salvaged materials in new construction, so the value of reused and recycled materials is expected to increase as this practice becomes more popular.

One way to quantify the economic benefits of DfAD is comparing the depreciation of a building designed without considering deconstruction and one in which it was prioritized. Depreciation can be defined as the reduction in value of an asset over time. The "straight line" method is one way to calculate depreciation, as shown below:

$$D_j = \frac{C - S_n}{n} \tag{7.3}$$

Where
    D_j = Depreciation in year j
    C = Cost
    S_n = Expected salvage value in year n
    n = Expected life of an asset

In the case of two structures, one designed for deconstruction and one for construction, the biggest anticipated difference would be the salvage value at the end of building life. A higher salvage value would decrease the annual depreciation of a building due to the ability of the owner to see a return on their investment through the selling of salvaged materials.

Webster provides discussion around feasible strategies structural engineers can incorporate into design procedures in order to achieve DfAD (Webster, 2010). First, he recommends that engineers design the building simply, using a standard layout with similar bay sizes throughout the structure in order to simplify the disassembly

TABLE 7.4
Material and Connection Selection Recommendations

Improve Adaptability	Simple connections with clear load paths
	Standard details and components
	Minimize number of members of different sizes
Ease Disassembly	Mechanical fasteners rather than adhesives or welding
	Steel and wood-framed structures rather than concrete
	Use salvaged materials when possible

of the building and sorting of materials which are salvaged. In addition to designing simply, prioritizing durability of the structure and components is important to minimize unforeseen or differential frequency of maintenance or repair. In addition to decreasing likelihood of repair during the original building life, choosing structural elements which will remain durable after deconstruction and have a useful life as a salvaged material is also beneficial.

Another strategy is the use of independent building systems, such as mechanical or building envelope systems. This can increase the ease of adaptation and deconstruction or even routine maintenance. During the life cycle of a building, especially one designed to be adaptable, mechanical systems may require multiple upgrades or replacements. Therefore, creating separation of this equipment from the structural components can improve access, decrease costs associated with maintenance or repair, and decrease the likelihood of damaging structural members in the process. This separation can be created using dedicated vertical chases or a raised floor.

Finally, a number of considerations associated with selection of materials and connections can improve the adaptability and deconstructability of a building, some of which are summarized in Table 7.4. Along a similar vein to designing simply, adaptability can be improved through the use of simple connections, standard details, and minimal different sizes of members. For example, the use of simple connections with clear load paths may make repurposing a space more straight-forward. Additionally, by minimizing the number of structural members of different sizes and using fewer, larger members rather than more, smaller members can decrease costs during initial construction as well as improve the ease of modifying a building layout. In terms of deconstructability, mechanical fasteners, steel or wood frames, and salvaged materials all ease disassembly and encourage repurposing of materials. Mechanical fasteners, for instance, are easier to remove from structural elements without causing damage than adhesives or welding.

Example Problem 7.3

A business owner looking to move his office into a new building has narrowed his search to two different properties which would meet his company's needs. Both buildings have an expected useful life of 50 years and an initial purchase price of

$500,000. Due to the simple layout, selection of standard connections and uniform structural members, and easily reused materials such as steel and wood framing, building 1 has an expected salvage value of $150,000. Building 2 features materials and connections which are much more difficult to salvage, such as adhesive connections and cast-in-place concrete elements, as well as a much more complex layout; therefore, the expected salvage value of building 2 is only $50,000. Use the straight-line depreciation equation to show the business owner the economic benefits associated with purchasing a building designed using DfAD.

- Using Equation 7.3 introduced in this chapter, calculate the depreciation of both buildings.

$$D_j = \frac{C - S_n}{n}$$

$$\text{Building 1: } D_j = \frac{\$500,000 - \$150,000}{50} = \$7,000$$

$$\text{Building 2: } D_j = \frac{\$500,000 - \$50,000}{50} = \$9,000$$

- Comparing the depreciation of the two buildings, the depreciation each year is $2,000 less if building 1, which was designed using the DfAD approach, is purchased.

7.4   CROSS LAMINATED TIMBER

At the start of the twentieth century, developments occurred which made the use of reinforced concrete economically viable enough to begin superseding traditional timber construction. Reinforced concrete is Portland Cement Concrete reinforced with steel for structural applications. As a result, timber as a construction material was reduced primarily to use in residential structures, light-weight construction, or particular structures desiring timber (Brandner *et al*, 2016). Unfortunately, cement production, which is required for reinforced concrete, and steel production each contribute to approximately 8% of global CO_2 emissions. As concerns regarding greenhouse gas emissions and climate change continue to grow, more sustainable building materials are being explored. Over the last 10 years, timber construction has begun to regain market share, due in part to the development of cross laminated timber (CLT), an innovative laminar timber product (Brandner *et al.*, 2016).

Cross laminated timber (CLT) is a prefabricated, solid engineered wood panel. CLT was developed in the 1990s as a result of the sawmill industry's need to identify a more lucrative use for side boards at the time (Guttmann, 2008). It is a lightweight construction material which also provides dimensional stability, strength, and rigidity along with superior acoustic, fire, seismic, and thermal performance (Brandner

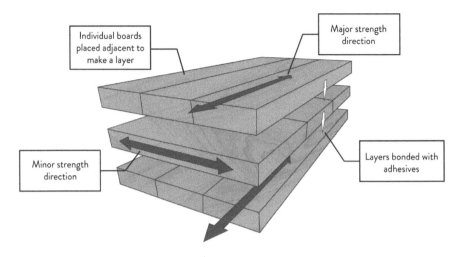

FIGURE 7.5   Cross laminated timber configuration (credit: M. Hufft)

et al., 2016). CLT is comprised of an uneven number of layers (typically 3, 5, or 7 layers), which are each made by placing individual boards adjacent to one another. The thickness and number of layers depend on the structural application for which the CLT is to be used. The layers are then arranged orthogonally and glued together under pressure using a polyurethane adhesive. Figure 7.5 displays a diagram of this structure. Similar to other engineered, layered wood products such as plywood, the in-plane dimensional stability is high due to minimized swelling and shrinkage from the cross layering. CLT differs from these previous wood products in its dimensions. The layers used to create CLT are thicker, giving the material larger dimensions both in-plane and out-of-plane, and increasing the versatility of CLT as a structural element. Strengths of CLT panels and laminations are specified in the major strength direction (direction of the grain of laminations in the outer layers of CLT panel), and minor strength direction (perpendicular to major strength direction).

CLT is gaining traction as a sustainable building material for a number of reasons. In terms of environmental benefits, as a timber product, CLT is made from the renewable resource of trees. As previously mentioned, production of both steel and concrete contributes significantly to global CO_2 emissions. CLT could decrease the amount of these building materials required for construction and possibly replace them entirely. Figure 7.6 displays a new dorm, Adohi Hall, constructed in the University of Arkansas campus which mitigated the use of steel and concrete through the use of timber construction, specifically CLT. Adohi Hall is the largest CLT building in the United States.

Decreasing the use or completely replacing steel and concrete would eliminate emissions released due to production of these materials. For example, a credit union office building in Hillsboro, Oregon, made of CLT was estimated to avoid 1,622 metric tons (1,788 US tons) of greenhouse gases which would have been emitted through steel manufacturing and construction (APA, 2018a). Although, due to a lack

FIGURE 7.6 Cross laminated timber residence hall at the University of Arkansas (Credit: C. Murray)

of data on life cycle analyses of CLT, there are concerns that the logging, manufacture, and transport associated with the production of CLT could result in a larger amount of CO_2 emissions than anticipated. Further research is needed on this subject to confidently support the claim CLT production yields fewer emissions than steel or concrete. Guo *et al.* performed a simulation analysis to compare the carbon emissions and energy consumption associated with heating and cooling a building during operation for a reinforced concrete building to that of a CLT building (Guo *et al.*, 2017). This study found the estimated energy consumption and carbon emissions in CLT buildings to be lower than a reinforced concrete building in all climate regions studied. CLT has the potential to reduce greenhouse gas emissions throughout operation of a structure, not only through initial material manufacturing. As more CLT structures are built, emissions and energy consumption throughout the life cycle of the building can be monitored in order to better understand the potential environmental benefits associated with CLT.

Economically, production and material costs of CLT structures can depend on the manufacturing scale; however, additional cost savings and economic benefits of CLT can be realized through shorter construction times and simplified assembly. CLT members are pre-fabricated, so they arrive on-site ready to be assembled. This has been shown to cut months off construction time as compared to steel and concrete construction times. Pre-fabrication also decreases the amount of material wasted as CLT panels are manufactured for specific end users (WoodWorks, 2012). Additionally, CLT construction can continue through cold weather, unlike concrete which has strict temperature tolerances. This can further reduce delays in construction and cut costs associated with delays. The previously mentioned credit union office building in Hillsboro, Oregon, achieved a 4-percent cost savings and was completed 4 months earlier than estimates for a comparable steel structure (APA, 2018a).

Social benefits of CLT are seen through creation of jobs, safer and quieter constructions, and aesthetically pleasing structures for people to use. As CLT continues to become more popular, additional labor will be needed in all stages of manufacturing in order to meet production demand. Because CLT is preplanned and prefabricated offsite, construction on-site can be accomplished faster, safer, and quieter. This benefits both the individuals assembling the CLT and those living or working in close proximity to the construction. Once the construction has been completed, CLT also extends social benefits to individuals throughout the use phase. CLT can provide valuable fire resistance due to the thick cross sections of panels charring slowly, protecting from rapid degradation in a fire. CLT buildings can also have fewer concealed spaces in a structure, reducing the likelihood of a fire spreading undetected (WoodWorks, 2012). Finally, the warmth and natural aesthetic created by timber construction can create a positive which promotes health and well-being for users of the structure (APA, 2018a).

CLT is manufactured using either kiln-dried lumber boards or structural composite lumber. These individual pieces of lumber which make up CLT are referred to as laminations. According to the Engineered Wood Association's (APA) Standard for Performance-Rated CLT, any softwood lumber species or species combination meeting minimum density requirements can be used in CLT manufacturing (APA, 2018b). The same lumber species or species combination must be used within a single layer; however, adjacent layers may be different species or species combinations. The thickness of laminations must be between 16 mm and 51 mm (5/8 inches and 2 inches). The moisture content of laminations at the time of CLT manufacturing is also specified as 12% for lumber and 8% for structural composite lumber. Different specifications exist for adhesives used to connect laminations. For instance, in the United States, adhesives are required to meet ANSI 405, *Standard for Adhesives for Use in Structural Glue Laminated Timber.*

Material characterization of CLT can be completed by either measuring mechanical properties of individual laminations or that of the constructed CLT element (Brandner *et al.*, 2016). Standardization of these requirements should be established moving forward in order to streamline design procedures. Currently, CLT is designed according to many different product-specific and technical approvals. Panels are manufactured for specific uses, to meet loading and span requirements. This creates flexibility in design. CLT has high axial load capacity for use in walls, is less susceptible to buckling, and has high shear strength to resist horizontal loads. Products are typically classified as "E" or "V." An "E" rating indicates machine stress–rated laminations, while a "V" rating indicates visually graded laminations. These ratings are used to characterize typical CLT layups, which are arrangements of layers in a CLT panel determined by the grade,

TABLE 7.5
Required Material Characteristics for Laminations in Typical CLT Layups (APA, 2018b)

CLT Layup	Laminations used in major strength direction					
	f_b (psi)	E (10^6 psi)	f_t (psi)	f_c (psi)	f_v (psi)	f_s (psi)
E1	4,095	1.7	2,885	3,420	425	140
E2	3,465	1.5	2,140	3,230	565	185
E3	2,520	1.2	1,260	2,660	346	115
E4	4,095	1.7	2,885	3,420	550	180
V1	1,890	1.6	1,205	2,565	565	185
V2	1,835	1.4	945	2,185	425	140
V3	1,575	1.4	945	2,375	550	180
CLT Layup	Laminations used in minor strength direction					
	f_b (psi)	E (10^6 psi)	f_t (psi)	f_c (psi)	f_v (psi)	f_s (psi)
E1	1,050	1.2	525	1,235	425	140
E2	1,100	1.4	680	1,470	565	185
E3	735	0.9	315	900	345	115
E4	945	1.3	525	1,375	550	180
V1	1,100	1.4	680	1,470	565	185
V2	1,050	1.2	525	1,235	425	140
V3	945	1.3	525	1,375	550	180

f_b = Characteristic bending strength of a lamination
E = Modulus of elasticity of a lamination
f_t = Characteristic axial tensile strength of a lamination
f_c = Characteristic axial compressive strength of a lamination
f_v = Characteristic shear strength of a lamination
f_s = Characteristic planar (rolling) shear strength of a lamination

number, orientation, and thickness of laminations. There are four grades of "E" CLT layups and three grades of "V" CLT layups. Table 7.5 lists the required material characteristics for laminations used in each of the typical layups, as specified by ANSI/APA PRG 320 CLT.

These minimum requirements are used along with section properties provided by the manufacturer in order to evaluate acceptance based on specified mechanical property testing. For instance, according to ANSI/APA PRG 320 CLT, bending tests are conducted in both the major and minor directions in accordance with the third-point load method of ASTM D198 or ASTM D4762. In this test, a lamination is simply supported and loaded at two points which are spaced at distances equivalent to one-third of the span length, as shown in Figure 7.7. The lamination is loaded at a constant rate to the point of failure. The bending strength is calculated using the peak load at failure.

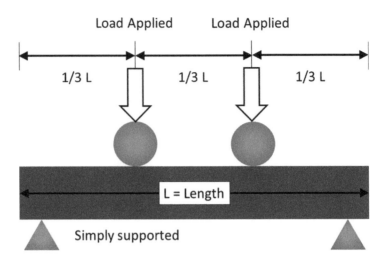

FIGURE 7.7   Third-point loading configuration (credit: s. Casillas)

A normal stress in a beam due to bending, σ_x, can be calculated using the following equation.

$$\sigma_x = \frac{-My}{I} \tag{7.4}$$

In this equation, M = the bending moment induced in the section
  y = the distance from the neutral axis to the point at which a beam is loaded
  I = the moment of inertia of the cross section

The negative sign of the moment indicates the beam is being loaded from the top and is therefore bending downward. Considering the third-point loading configuration described above for a lamination of CLT, a theoretical value of σ_x could be calculated at both points of loading.

Example Problem 7.4

A 10-foot long lamination to be used in a CLT layup is subjected to a uniform load of w = 30 lb/ft in the major strength direction. This lamination has a height of h = 2 inches and width of b = 2 inches. Assume the lamination is being evaluated as a simply supported beam. Calculate the normal stress due to bending induced in the major strength direction for this lamination. In terms of bending strength alone, which "E" CLT layups would provide adequate bending strength to withstand this loading?

- Begin by calculating the bending moment, M. Because this is a simply supported beam subjected to a uniform load, the following equation is used to calculate moment.

$$M = \frac{wL^2}{8}$$

Where w is the uniform load applied and L is the length of the lamination.

$$M = \frac{30\ \text{lb/ft} \times 10\ \text{ft}^2}{8} = 375\ \text{lb} \cdot \text{ft}$$

- Next, calculate the moment of inertia for this lamination.

$$I = \frac{bh^3}{12} = \frac{2^4}{12} = 1.33\ \text{in}^4$$

- Because this lamination has a square cross section, the distance to the neutral axis will be half the height of the lamination. Therefore, y = 1 inch.
- Finally, calculate the normal stress due to bending:

$$\sigma_x = \frac{-My}{I} = \frac{375\ \text{lb} \cdot \text{ft} \times 1\ \text{in}}{1.33\ \text{in}^4} \times \frac{12\ \text{in}}{1\ \text{ft}} = 3,375\ \text{psi}$$

- Considering the bending strengths for laminations used in the major strength direction shown in Table 7.4, this bending stress is only less than the strength values given for E1 and E2 CLT layups. Therefore, an E1 or E2 CLT layup would provide adequate bending strength to withstand this loading.

HOMEWORK PROBLEMS

1. A local contractor is investigating replacing 50% virgin Portland cement with fly ash. The concrete with 100% concrete has a compressive strength of 25 MPa at 3 days and a compressive strength of 50 MPa at 90 days. The concrete with 50% fly ash has a compressive strength of 22 MPa at 3 days and 55 mPa at 90 days. Using the common relationship between compressive strength and modulus of elasticity, calculate the secant modulus of elasticity (E_c) at these 4 strength levels.

2. Using the fly ash replacement levels in Problem 1, along with data comparing the use of fly ash versus Portland cement, prepare a summary of the economic and environmental benefits of using fly ash versus Portland cement. Discuss your findings using the format provided under the sidebar in Chapter 1 "Writing a High-Quality Essay."

3. A firm in central Illinois would like to explore using bamboo as a substitute for steel for scaffolding. However, they are concerned about the load-carrying ability of the bamboo versus the steel. Assume that both materials have a fixed connection at each end, the length of the column will be 20 ft, and the radius of gyration of each material will be 1.15 inches. Also, assume that the modulus of elasticity of the steel is 25.8×10^6 psi. Compare the steel

to bamboo in a wet state ($\alpha = 2.35$, $E = 1.25 \times 10^6$ psi) and in a dry state ($\alpha = 2.35$, $E = 1.75 \times 10^6$ psi).

4. A firm in Nanjing, China, would like to explore using steel as a substitute for bamboo for scaffolding. However, they are concerned about the load-carrying ability of steel versus bamboo. Assume that both materials have one end fixed and the other end pinned, the length of the column will be 15 ft, and the radius of gyration of each material will be 1.20 inches. Also, assume that the modulus of elasticity of the steel is 26.7×10^6 psi. Compare the steel to bamboo in a wet state ($\alpha = 2.35$, $E = 1.15 \times 10^6$ psi) and in a dry state ($\alpha = 2.35$, $E = 1.65 \times 10^6$ psi).

5. As the structural engineer designing a new multi-tenant space you need to justify higher initial costs associated with DfAD to the owner. Recalling the concept of life cycle cost analysis (LCCA), calculate the net present value (NPV) of the space if initial construction cost is $2,000,000, maintenance costs $25,000 every 10 years, modifying the space for new tenants costs $2,000 every two years, and the salvage value at the end of life is $500,000. The analysis period is 40 years, and the discount rate is 2.5%.

6. After presenting the owner with the NPV calculated in Problem 5, he is still not fully bought in to prioritizing designing with deconstruction in mind. Therefore, to illustrate your point, calculate the depreciation of the building using your projected initial cost of $2,000,000 and salvage value of $500,000 over the 40-year period compared to an alternative with a lower initial cost of $1,650,000 and salvage value of $50,000 at the end of the 40-year period.

7. A 12-foot long lamination to be used in a CLT layup is subjected to a uniform load of w = 55 lb/ft in the major strength direction. This lamination has a height of h = 5 inches and width of b = 2 inches. Assume the lamination is being evaluated as a simply supported beam. Calculate the normal stress due to bending induced in the major strength direction for this lamination. In terms of bending strength alone, which "V" CLT layups would provide adequate bending strength to withstand this loading?

8. If the same lamination described in Problem 7 is subjected to a uniform load of w = 15 lb/ft in the minor strength direction, calculate the normal stress due to bending induced in the minor strength direction for this lamination. Would the same "V" CLT layups provide adequate strength? If not, which "V" CLT layups would you recommend?

REFERENCES

Ahlman, A., Edil, T., Natarajan, B., Ponte, K. System Wide Life Cycle Benefits of Fly Ash. *2015 World of Coal Ash (WOCA) Conference*, Nashville, TN, May 5–7, 2015.

APA – The Engineered Wood Association. APA Case Study – First Tech Federal Credit Union: The Building that Wanted to Be Mass Timber. *Form No. U115*, 2018a.

APA – The Engineered Wood Association. Standard for Performance-Rated Cross-Laminated Timber (ANSI/APA PRG 320), 2018b.

Athena Sustainable Materials Institute. Minnesota Demolition Survey: Phase Two Report, Prepared for Forintek Canada Corp., 2004a.

Athena Sustainable Materials Institute. Athena Software Version 3.01, Database Version 2.00, 2004b.

Brandner, R., Flatscher, G., Ringhofer, A., Schickhofer, G., Thiel, A. Cross Laminated Timber (CLT): Overview and Development. *European Journal of Wood and Wood Products*, Published online, 74(3), pp. 331–351, 2016.

EPA. *Climate Change Mitigation Strategies in the Forest and Agricultural Sectors.* U.S. Environmental Protection Agency (EPA), Climate Change Division, Washington, DC, June, 1995.

Guttmann, E. Brettsperrholz: Ein Produktporträt [Cross lami- nated timber: a product profile]. *Zuschnitt*, 2008, 31, 12–14 (in German).

Guo, H., Liu, Y., Chang, W., Shao, Y., Sun, C. Energy Saving and Carbon Reduction in the Operation Stage of Cross Laminated Timber Residential Buildings in China. *Sustainability*, 2017, 9(2), pp. 1–17.

ICF International. *Documentation for Greenhouse Gas Emission and Energy Factors Used in the Waste Reduction Model (Warm).* U.S. Environmental Protection Agency Office of Resource Conservation and Recovery, 2016.

Langley, W., Carette, G., Malhotra, V. Strength Development and Temperature Rise in Large Concrete Blocks Containing High Volumes of Low-Calcium (ASTM Class F) Fly Ash. *ACI Materials Journal*, 1992, 89(4), 362–368.

Lippiatt, B., Ahmad, S. Measuring the life-cycle environmental and economic performance of concrete: the bees approach. *Proceedings of the International Workshop on Sustainable Development and Concrete Technology*, Beijing, China, May 20–21, 2004, 213–230.

Lu, Y. An Investigation of High-Volume Fly Ash Concrete for Pavements. M.S. Thesis, University of Queensland, 2007.

Moffatt, S., Russell, P. Assessing the Adaptability of Buildings. *IEA Annex 31 Energy-Related Environmental Impact of Buildings*, 2001.

Ondova, M., Estokova, A. LCA and Multi-Criteria Analysis of Fly Ash Concrete Pavements. *International Journal of Environmental, Chemical, Ecological, Geological and Geophysical Engineering*, 2014, 8(5), 320–325.

Santero, N., Loijos, A., Akbarian, M., Ochsendorf, J. Methods, Impacts, and Opportunities in the Concrete Pavement Life Cycle. Concrete Sustainability Hub, Massachusetts Institute of Technology, August, 2011.

Sharma, B., Gatoo, A., Bock, M., Ramage, M. Engineered Bamboo for Structural Applications. *Construction and Building Materials*, 81, 2015, 66–73.

Thomas, M. *Optimizing the Use of Fly Ash in Concrete.* IS548, Portland Cement Association, Skokie, IL, 2007.

Webster, M. Structural Design for Adaptability and Deconstruction: A Strategy for Closing the Materials Loop and Increasing Building Value. *2007 Structures Congress: New Horizons and Better Practices. American Society of Civil Engineers*, 2007.

Webster, M. Design for Adaptability and Deconstruction. *Sustainability Guidelines for the Structural Engineer.* American Society of Civil Engineers, 2010, 85–92.

Widenoja, R. Sub-Optimal Equilibriums in the Carbon Forestry Game: Why Bamboo Should Win But Will Not. Masters thesis, Fletcher School of Law and Diplomacy, Tufts University, April, 2007.

WoodWorks Solid Advantages. Information Sheet WW-012, 2012.

8 Application
Transportation Sustainability

It is the mark of an educated mind to be able to entertain a thought without accepting it.

Aristotle

The Transportation and Development Institute (T&DI) is one of the specialty institutes of the American Society of Civil Engineers. Like many institutes, T&DI has a host of committees that track further areas of interest for the institute. The committees of T&DI address topics such as aviation, freight and logistics, infrastructure systems, rail and public transit, roadways, and development. As a part of the development council, T&DI has a committee on sustainability and the environment. This committee on sustainability addresses all facets of sustainability within transportation engineering. This chapter will cover four of those areas to provide a glimpse into how sustainability can be applied in the fourth and final application area of sustainability, which is transportation:

1. Material reuse – RAP and RAS
2. Mutlimodal transportation
3. Intelligent transportation systems
4. Crash modification factors

8.1 MATERIAL REUSE – RAP AND RAS

Asphalt mixtures are made of both aggregate and asphalt binder. The aggregate, which is approximately 93–97% of the asphalt mixture by weight, is designed to provide a skeleton to carry the weight of vehicles passing over the pavement. The asphalt binder, 3–7% of the asphalt mixture by weight, is designed to hold the aggregate together and to provide flexibility to the pavement structure during traffic loading and ground movement. However, over time, the asphalt pavement loses its flexibility due to natural weathering. Weathering can include oxidation from the sun and wind, moisture damage, and traffic damage. This weathering is usually confined to the upper layer of the asphalt pavement structure and does not typically extend more than 1.5–2.0 inches into the pavement. Therefore, it is common to mill off the existing surface course of a pavement and replace it with a new asphalt mixture. The material that was milled off the existing surface course of the pavement is called Reclaimed Asphalt Pavement, or RAP. RAP, though weathered, still contains properties that could be beneficial to asphalt concrete. According to the National Asphalt Pavement Association (NAPA), over 68.3 million tons of RAP was used in asphalt mixtures in 2012 (Hansen and Copeland, 2013). This RAP replaces not only the aggregate, but also the asphalt binder partially. The use of RAP is quite well established in asphalt mixtures. The National Cooperative Highway Research Program (NCHRP) performed an extensive study examining the mix design, performance, and materials management for RAP in asphalt mixtures (West et al., 2013). The report states that the use of RAP, even in high quantities, is perfectly acceptable with the proper design process, but good management practices are essential for proper performance. Figure 8.1 shows a front-end loader collecting load of RAP during asphalt mixture production.

Another material that can supplement asphalt binder in asphalt mixtures is Recycled Asphalt Shingles, or RAS. Asphalt shingles can come from either scrap during manufacturing, or they are ripped off roofs after their service life on buildings is over. After some processing, RAS can be incorporated into asphalt mixtures. In 2012, NAPA estimated 1.9 million tons of RAS in asphalt mixtures (Hansen and Copeland, 2013). While RAS does not replace aggregate in asphalt mixtures, using RAS in U.S. roadways in 2012 conserved approximately 2.1 million barrels of asphalt binder, saving approximately $228 million. Even more material and economic savings were seen for RAP, which conserved over 19 million barrels of asphalt binder in U.S. roadways, saving approximately $2 billion. While the use of RAS in asphalt mixtures is not quite as prevalent as RAP, there is still much research indicating that, again, with proper design and management practice of the material, RAS can

FIGURE 8.1   A stockpile of RAP in northwest Arkansas (credit: A Braham)

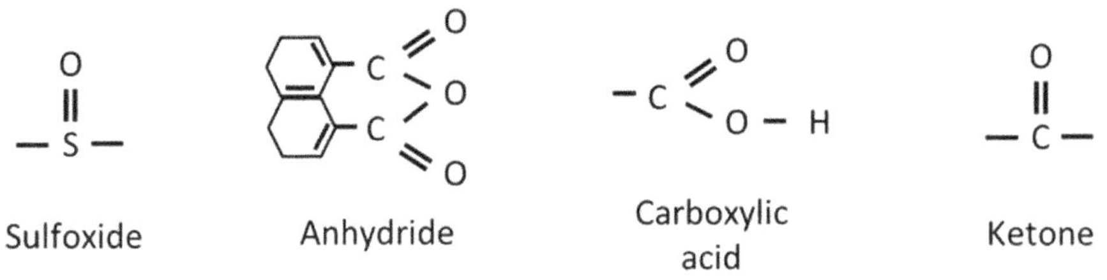

FIGURE 8.2   Oxidative hardening of four asphalt binder molecules (credit: A Braham)

be successfully utilized in asphalt mixtures (Zhou *et al.*, 2013; Ozer *et al.*, 2013; Cooper *et al.*, 2014).

While both RAP and RAS have been proven to be a quality substitute for virgin material, the binder portion of the RAP and RAS tends to oxidize, which makes it stiffer. RAP and RAS are both petroleum based, but during their in-service life they are exposed to oxygen, and the asphalt binder goes through what is called oxidative hardening. In short, oxidative hardening occurs when polar, oxygen-containing chemical groups are introduced to asphalt molecules. More detail on oxidative hardening and other forms of hardening can be found in Peterson, 2009. Four examples of oxidative hardening on the molecular scale are shown in Figure 8.2.

The hardening of asphalt binder influences the classification of the binder. In the United States, the Superpave system is used to classify binder, and binder is classified by "grading" the binder. The Performance Grade (PG), or PG binder grading system, has both high and low temperatures. The high temperature of the PG grade is related to the anticipated high air temperature that the road will be exposed to. Improperly designed roads may rut under higher temperatures. The low temperature of the PG grade is related to the anticipated low pavement surface temperature that the road will be exposed to. Improperly designed roads may crack under low temperatures. A stiffer binder increases the high temperature binder grade and also increases the lower temperature binder grade. In theory, this makes the mixture less susceptible to rutting, but more susceptible to cracking. Therefore, care must be taken to ensure that the mixture's binder grade stays within the proper design range, which is one of the recommendations from research for using both RAP and RAS.

There are several challenges associated with using RAP and RAS in asphalt mixtures. The largest challenge is providing a consistent product. RAP is obtained from milling an existing roadway. A common rehabilitation strategy for asphalt roadways is to mill 2 inches off the top of a pavement surface and then lay down 2 inches of new asphalt mixture. This eliminates any surface distresses and provides a new traveling surface. However, after a road is milled, the material is generally taken back to an asphalt plant, where it is crushed and mixed with other millings. The millings are obtained from many different roads, all of which may have had different original aggregate and asphalt binders, and that which likely are of different ages. This means that RAP stock piles are a very diverse pile of materials; hence, it can prove difficult to ensure a consistent product going into the asphalt mixture. RAS can be even more diverse, as there are two primary types of RAS: waste from asphalt shingle production and tear-off shingles. Waste from the production of asphalt shingles is preferred, as the material is not exposed to weathering or oxidation. However, tear-off shingles, shingles that are removed from roofs, are highly oxidized, and also highly variable. While consistent RAP and RAS products are a challenge to produce, if properly managed with suppliers, they can be successfully incorporated into a roadway as a useful recycled material.

Example Problem 8.1

On the website "pavementinterative.org" there is a section on Superpave Performance Grading. About a half of the way down the page is Figure 2 – PG binder specification taken from NAPA. Using this chart, record the testing temperatures for each test when evaluating a PG64-22 binder. In addition, if you add a stiffer material to an asphalt mixture (such as RAP or RAS), what do you anticipate would happen to the low (-22°C) and high binder (+64°C) temperatures?

Going through the chart, the test temperatures are within each box of the table. For a PG64-22, the following tests require the following test temperatures:

- Original binder:
  - Flash point: 230°C
  - Viscosity: 135°C
  - Dynamic shear: 64°C
- Rolling Thin Film Oven (RTFO) residue:
  - Dynamic shear: 64°C
- Pressure Aging Vessel (PAV) residue:
  - Dynamic shear: 25°C
  - Creep stiffness: -12°C
  - Direct tension: -12°C

In regard to the high and low binder temperatures, if you add RAP and RAS to an asphalt mixture, it will make the mixture stiffer, thereby increasing the high and low temperature binder grades. So, for example, if you add RAP or RAS to an asphalt mixture with a PG64-22 binder, adding enough RAP or RAS may increase the binder grade to a PG70-16.

8.2   MULTIMODAL TRANSPORTATION

When considering modes of transportation, two perspectives should be examined: passenger travel and freight movement. In terms of passenger travel, travel by car versus bus versus self-propelled travel (for example, bicycles), is frequent. For example, the European Cyclist Federation looked at the impact of biking, taking a bus, and driving a car. By examining the fuel CO_2 emissions, they found that driving produced 0.81 pounds/mile (assuming an occupancy of 1.16 passengers per vehicle), taking a bus produced 0.34 pounds/mile (assuming an occupancy of 10 passengers per bus), while biking produced 0.6 pounds/mile (assuming one person per bicycle).

FIGURE 8.3   Bicycle versus car versus bus vehicle density (credit: Carlton Reid)

The CO_2 emissions for cars and buses was calculated for emissions linked to production, distribution, and consumption of fuel, while the CO_2 emissions for biking was calculated by food production for the bicyclist. Another benefit of utilizing mass transit and bicycles is the reduction of vehicle volume on the roadway. A famous picture from the City of Munster, Germany, (Figure 8.3) shows the density of vehicles on a city street for 60 people.

It is clear that the 60 bicycles and 1 bus take up far less space on the roadway versus the 60 individual vehicles. In essence, reducing the density of vehicles on the roadway generally increases the flow of traffic and enhances the Level of Service. However, many of these traditional studies assume the use of conventionally powered and individually owned vehicles. In addition, the studies have been very one dimensional, looking at either just the economic or environmental impact of transit. More recent studies have investigated hybrid cars and buses and have incorporated concepts of ride-sharing into their analyses. In addition, some recent studies have expanded the scope of their analyses to include more sustainability metrics. For example, a study in 2014 (Mitropoulos and Prevedouros, 2014) looked at conventional internal combustion cars versus hybrid electric cars, both with and without car sharing, and ranked these four groups versus a traditional diesel bus and a hybrid diesel-electric bus. These six vehicle configurations were explored across five sustainability concepts:

1. Minimize environmental impacts
2. Minimize energy consumption
3. Maximize and support a vibrant economy
4. Maximize user and community satisfaction
5. Maximize technology performance to help a community meet its needs

While many assumptions went into the analysis, it was found that when looking at both Passenger Miles Traveled (PMT) and Vehicle Miles Traveled (VMT), the car sharing and hybrid cars performed better than the buses.

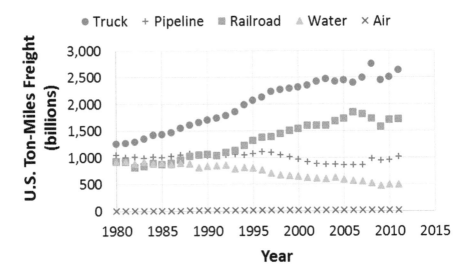

FIGURE 8.4   Freight shipping by mode in the United States (credit: A Braham)

Along with passenger mobility, another subject to consider is freight, which must also be transported around the country. A convenience that freight has versus passengers is that travel time is often less restricted and factors such as comfort do not need to be addressed. Assuming that the perishable or time-sensitive material is not being hauled, there are five primary modes of freight transportation: waterways, rails, highways, air, and pipelines. However, the majority of freight in the United States is generally shipped by either railroad or truck. This data is shown in Figure 8.4, which was compiled from data available online from the Bureau of Transportation Statistics.

Studies have shown that for a typical barge movement of freight on water, the same CO_2 emissions would be generated for hauling the same ton-miles of freight on 25 train cars or 297 trucks on the highway. Costs are also less for movement on water, as rail was found to be about 1.7 times higher than ship, while trucks were found to be 2.8 times higher (Cenek *et al.*, 2012). Other studies have seen even higher cost differences, with trucks costing five to ten times more per ton-mile versus rail (Rasul, 2014). In this study, Rasul determined that it costs an estimated $0.10–$0.20 per ton-mile to ship by truck yet only $0.01–$0.04 per ton-mile by rail.

In addition to the potential environmental and economic savings of shipping freight by rail and ship versus truck, highway pavement materials could potentially be reduced as well. One of critical inputs into pavement design is the number of equivalent single axle loads, or ESALs. In general, higher ESALs mean that a thicker pavement structure is required, increasing the economic and environmental impact of the roadway. ESALs can be computed using Equation 8.1:

$$ESAL_i = f_d \times G_m \times AADT_i \times 365 \frac{days}{year} \times N_i \times F_{Ei} \qquad (8.1)$$

where

$ESAL_i$ = equivalent accumulated 18,000-lb single-axle load for the axle category i per year

f_d = design lane factor (percent truck volume on design lane)

G_{rn} = growth factor for a given growth rate r and design period $n = [(1+r)^n - 1]/r$

$AADT_i$ = first-year annual average daily traffic for axle category i

N_i = number of axles on each vehicle in category i

F_{Ei} = load equivalency factor for axle category i

The design lane factor (f_d), the growth factor (G_{rn}), the first-year annual average daily traffic for axle category i ($AADT_i$), and the number of axles on each vehicle (N_i) are all relatively straight forward, and usually given, yet, the load equivalency factor (F_{Ei}) is a critical input. Tables are available in the 1993 AASHTO pavement design guide that relate the terminal serviceability index and the pavement structural number to the load equivalency factor. For example, one table is provided in the Fundamentals of Engineering (FE) reference manual for the load equivalency factor. The table is built from a roadway that has a terminal serviceability index of 2.5 and a pavement structural number of 5.0. In the table, the load equivalency factors for a passenger car (1,000 lb or 4.45 kN) is only 0.00002, whereas the standard semi-truck weight for a single axle is 18,000 lb (80.0 kN), with a load equivalency factor of 1.0. This means that a standard loaded semi-truck has approximately 50,000 times more ESAL influence on a road compared to a passenger car. This clearly shows that trucks are what deteriorate pavements, so agencies must design for the "worst-case" scenario, which are the trucks. By shifting freight movement from highways to either rail, water, or pipelines, the life expectancy of pavements would be extended.

Example Problem 8.2

A six-lane interstate is being built through Fayetteville, Springdale, Rogers, and Bentonville in Northwest Arkansas. Traffic volume forecasts estimate that there will be 62,000 average annual daily traffic (AADT) in both directions during the first year of operation. The following vehicle mix is expected:

1. Passenger cars (2000 lbs/axle, $F_{Ei} = 0.0002$) = 60%
2. 3-axle single-unit trucks (10,000 lb/axle, $F_{Ei} = 0.0877$) = 25%
3. 5-axle tandem-unit trucks (20,000 lb/axle, $F_{Ei} = 0.1206$) = 15%

If the growth factor is anticipated to be 33.06 (from an annual growth rate of 5% and a 20-year design period), and the percent truck volume on the design lane is 35%, compute the total ESALs for the roadway.

To solve the problem, Equation 8.1 must be solved for each vehicle class. Passenger cars (pc):

$$ESAL_{pc} = f_d \times G_{rn} \times AADT_i \times 365 \times N_i \times F_{Ei}$$

$$= 0.35 \times 33.06 \times 62,000 \times 365 \times 0.60 \times 2 \times 0.0002 = .063 \times 10^6$$

3-axle single-unit trucks (3t):

$$ESAL_{3t} = 0.35 \times 33.06 \times 62,000 \times 365 \times 0.25 \times 3 \times 0.0877 = 17.2 \times 10^6$$

5-axle tandem-unit trucks (5t):

$$ESAL_{5t} = 0.35 \times 33.06 \times 62,000 \times 365 \times 0.15 \times 5 \times 0.1206 = 23.7 \times 10^6$$

$$\text{Total ESALs} = ESAL_{pc} + ESAL_{3t} + ESAL_{5t}$$
$$= 0.63 \times 10^6 + 17.2 \times 10^6 + 23.7 \times 10^6 = \underline{41.0 \times 10^6}$$

It is interesting to note that the passenger cars, while taking up 60% of the traffic stream, have essentially a negligible impact on the ESAL count, and hence, have a negligible impact on the deterioration of the roadway structure as well.

8.3   INTELLIGENT TRANSPORTATION SYSTEMS

In 2013, the National Academies and the Executive Committee of the Transportation Research Board published a discussion of six critical issues in transportation (TRB, 2013). In the first paragraph of the first page of text, there is a discussion about the lost time in traffic congestion. Just a couple of paragraphs later, the second critical issue discusses road safety, and the third critical issue identifies transportation's unsustainable impact on the environment. These three points, interestingly, are almost identical to comments made by Ban Ki-moon in the UN's sustainable mobility discussion (UN, 2013). Ki-moon's comments revolved around the need to improve road safety, reduce congestion for people and freight, and minimize environmental impact of transportation systems. One potential solution to addressing these three comments and the discussion of critical issues is Intelligent Transportation Systems, or ITS.

ITS collect, store, analyze, and distribute data on the movement of people and freight. ITS provide real-time travel information services and provide management models across all modes of transportation. ITS can help with incident/crash management, emergency response systems, crash prevention, and roadway maintenance systems. Some examples of ITS for drivers include variable message signs along the side of the road, on-board vehicle systems (such as lane departure warning systems), advanced emergency braking systems, and on-board diagnostics. Other types of ITS resources for drivers are apps, such as IDrive Arkansas, which can provide construction, weather, and traffic information in real time to travelers on their cell phones. An example of a variable message sign in Missouri is shown in Figure 8.5. While the sign in Figure 8.5 did not display any emergency information at the time of the

FIGURE 8.5   ITS variable message sign in Missouri (credit: A Braham)

picture, it can quickly be changed from a remote location to warn drivers of issues coming up on the roadway. These signs also can be used for emergency messages such as AMBER (America's Missing: Broadcast Emergency Response) alerts.

A specific application that can be explored related to ITS is the effect of driver reaction time on stopping sight distance. The stopping sight distance (SSD) is a combination of the distance associated with driver reaction time (D_{rt}) and braking distance (D_b), which can be shown in Equation 8.2:

$$SSD = D_{rt} + D_b = 1.47Vt + \frac{V^2}{30\left(\left(\dfrac{a}{32.2}\right) \pm G\right)} \tag{8.2}$$

where
  V = design speed (mph)
  t   = driver reaction time (sec)
  a   = deceleration rate (assume AASHTO recommended value of 11.2 ft/sec^2)
  G = percent grade divided by 100 (will be in decimal form)

One study on reaction time (Porter *et al.*, 2008) examined the effect of sound warnings on braking distance in younger male drivers and older male drivers. The young males were 30–50 years old, and the old males were 70+. Sound warnings included auditory alerts 100 m before crosswalks, school zones, playgrounds, red light cameras, and deer crossings, in addition to the traditional visual signage from the Manual of Uniform Traffic Control Devices (MUTCD). Four scenarios were examined with the two age groups: expected events with and without sound warnings (expected meaning a person crossing at a crosswalk, a light changing from green to red on approach, etc.) and unexpected events with and without sounds warnings (unexpected meaning a car entering the roadway just after a school zone alert, a pedestrian in the street not at a crosswalk, etc.). Some general characteristics were examined to determine statistical differences and similarities between younger and older drivers:

- Differences
  - Age – the age difference between the older and younger drivers was statistically significant
  - Driving experience – older drivers had more driving experience
  - Number of days driven in past week – older drivers had driven more in past week
  - Leg strength – younger drivers had higher leg strength
  - Mobility – younger drivers had higher mobility
  - Visual information processing – younger drivers processed information more quickly
- Similarities
  - Visualizing missing information – both age groups were able to visualize missing information in a similar manner

Using a driving simulator, 16 younger drivers and 14 older drivers were put through multiple driving situations that explored the expected and unexpected events, with and without sound warnings. Table 8.1 summarizes the average reaction times, along

TABLE 8.1
Effect of Sound Warnings on Reaction Times and Distances

Age	Event	Alert	Average reaction time (sec)	Average reaction distance (ft)
Younger	Expected	Sound warning	1.8	98
		No sound warning	2.2	120
	Unexpected	Sound warning	2.5	137
		No sound warning	2.4	131
Older	Expected	Sound warning	2.3	126
		No sound warning	3.1	170
	Unexpected	Sound warning	2.5	137
		No sound warning	2.8	153

with the corresponding reaction time distances, with the design speed of the simulations set at 37.2 mph.

Table 8.1 clearly demonstrates that older drivers had longer reaction distances versus younger drivers and that both sets of age groups had longer reaction distances when events were unexpected versus expected. In general, the sound warning also assisted in reducing reaction distances, except with younger drivers in an unexpected event, where the distances were essentially the same. This is one example of how an ITS technology onboard a vehicle can reduce reaction distances, and thus SSD, creating safer environments on the roadway. It is interesting to note that the default value for reaction time used by AASHTO is 2.5 seconds, which comes from the combination of 1.5 seconds to perceive the need to apply brakes and 1.0 seconds to begin the braking process. This 2.5 seconds falls at the very long end of the younger drivers in Porter's study, while it falls quite near the short end for the older drivers.

Example Problem 8.3

Determine the difference in stopping sight distance for an older driver between receiving a sound warning and not receiving a sound warning with an expected event. Assume a speed of 72 mph, on a downward grade of 2.3%.

In order to solve this problem, Equation 8.2 is utilized. Recall that the standard AASHTO value for deceleration rate can be used (11.2 ft/s²) and that the grade must be in decimal form.

With sound warning:

$$SSD = 1.47Vt + \frac{V^2}{30\left(\left(\dfrac{a}{32.2}\right) \pm G\right)}$$

$$= 1.47 \times 72 \times 2.3 + \frac{72^2}{30\left(\left(\dfrac{11.2}{32.2}\right) - 0.025\right)} = 741\,\text{ft}$$

Without sound warning:

$$SSD = 1.47Vt + \frac{V^2}{30\left(\left(\dfrac{a}{32.2}\right) \pm G\right)}$$

$$= 1.47 \times 72 \times 3.1 + \frac{72^2}{30\left(\left(\dfrac{11.2}{32.2}\right) - 0.025\right)} = 826\,\text{ft}$$

Therefore, the difference for older drivers with and without a sound warning is 826 − 741 = 85 ft

8.4   CRASH MODIFICATION FACTORS

The discussion on transportation applications of sustainability has so far covered material reuse (RAP and RAS), multimodal transportation (passenger travel by car, bus, bicycle; freight travel by truck, rail, water, pipeline, and air), and intelligent transportation systems (ITS). These concepts all had components of the economic, environmental, and social pillars of sustainability. Crash Modification Factors (CMFs) are a bit unique; however, as they revolve primarily around the social pillar of sustainability (with the concept of safety) and secondly around the economic pillar (litigation associated with crashes, injury, and even death), with CMFs, there is less to be said about environmental impacts.

The concept of safety has become so important in the United States that, in 2010, the first edition of the Highway Safety Manual was released by AASHTO (AASHTO, 2010). This manual is similar to other AASHTO publications, such as the Manual for Bridge Element Inspection or A Policy on Geometric Design of Highways and Streets (the green book), but AASTHO's Highway Safety Manual focuses exclusively on safety. Two of the largest subjects covered in the book are Safety Performance Functions and CMFs. In the United States, according to the National Highway Traffic Safety Administration (NHTSA), there were over 32,000 traffic crash fatalities in 2014. For the purpose of giving an important glimpse into the social and economic pillars of sustainability, as related to the field of transportation engineering, this chapter will focus exclusively on CMFs, which are implemented in the AASHTO Highway Safety Manual on the following roadway types:

- Roadway segments
- Intersections
- Interchanges
- Special facilities and geometrics situations
- Road networks

In order to discuss CMFs, a general overview of vehicle crash considerations is helpful. Broadly speaking, there are four generally categorized factors involved with crashes: human factors, vehicle conditions, roadway conditions, and the environment. Human factors revolve around the driver and the driver's actions. For example, younger drivers have less experience driving and therefore may make unexpected decisions while driving, whereas older drivers' reaction times increase (as seen in Chapter 8.3). External factors that engineers can control, however, also play a factor. For example, crashes can occur as a result of information overload to drivers, through roadway design and signage. Therefore, it is important that information is provided in an orderly and consistent way. Avoiding information overload to drivers can be achieved, for instance, by placing a series of signs that give information in a progressive, orderly manner, which helps drivers rank the importance of information. These concepts are discussed in more detail in the Manual of Uniform Traffic

Control Devices (MUTCD). The MUTCD can be downloaded for free from FHWA's website; simply google "Manual of Uniform Traffic Control Devices" and FHWA's website with the .pdf download should be one of the first links.

The second factor of crashes is the vehicle condition. Obviously, if a car has bad breaks or the steering is not fully responsive, there is an increased likelihood of a crash. In addition, as technologies progress, technologies such as power steering and anti-lock brakes have reduced the number of crashes, while technologies such as seat belts and airbags have reduced the amount of injuries and fatalities associated with crashes. As these technologies degrade over time in a vehicle, the number of crashes and injuries increases.

The third factor of crashes is the roadway condition. Conditions of the roadway can be broken down into four parts: the pavement, the shoulders, intersections, and the traffic control system. When considering the pavement, there must be enough surface friction between the roadway and the tires so that drivers can maintain control of the vehicle, while wide shoulders give space for disabled vehicles to move off active lanes of traffic. Intersections must be properly designed for easy lines of sight for drivers to observe cars approaching from different directions. Finally, components of the traffic control system, such as stop lights, must be easily visible as vehicles approach and waiting at the intersection.

The fourth and last factor generally recognized about crashes is the environment. Weather, for instance, plays a significant role in crashes. Sitting water on the pavement surface can cause hydroplaning or can freeze and form ice, both of which result in a loss of friction between the tire and pavement surface. Fog can reduce the visibility of the driver, which reduces the stopping sight distance. Besides weather, another environmental factor that is not commonly addressed is the level of lighting. It is estimated that the number of fatal crashes during daylight is about the same as the number that occur in darkness, but only 25% of the vehicle-miles traveled occur at night (Lutkevich et al., 2012). This means that fatal crashes are three times as likely to occur when it is dark versus when it is light. While transportation engineers have only partial control over these four factors of crashes, the roadway infrastructure should be designed in ways that minimize the external factors of crashes.

With the dangerous and significant ramifications of crashes, two methods of quantifying crash rates have been developed, one for intersections and one for roadways. For intersections, crash rates are presented as Crash Rate per million Entering Vehicles (RMEV), which is calculated by Equation 8.3:

$$RMEV = \frac{A \times 1,000,000}{V} \qquad (8.3)$$

where
  A = number of crashes, total/type occurring in a single year
  V = average daily traffic (ADT) entering the intersection × 365 days/year.

A similar equation has been developed for roadway segments. For roadway segments, crash rates are presented as Crash Rate per hundred million vehicle miles (RMVM), which is calculated by Equation 8.4:

$$RMVM = \frac{A \times 100,000,000}{VMT} \qquad (8.4)$$

where

A   = number of crashes, total/type occurring in a single year
VMT = vehicle miles of travel during given period
      = average daily traffic (ADT) on roadway segment × number of days in study period × length of road.

Using these two calculations, an examination of CMFs can begin.

In this chapter, roadway segments is the only type of roadway that will be examined in the AASHTO Highway Safety Manual. Roadway segments consist of the following types of roads:

- Rural
  - Two-lane road
  - Multiplane highway
  - Frontage road
- Freeway
- Expressway
- Urban arterial
- Suburban arterial

Each of these road types has various treatments, also called countermeasures, associated with reducing crashes, including modifying the lane width, adding lanes by narrowing existing lanes and shoulders, removing lanes, adding/widening a paved shoulder, providing a raised median, changing the width of the existing median, and increasing the median width. These treatments can then be associated with CMFs, from which the number of crashes prevented can be calculated. In order to calculate the number of crashes prevented, Equation 8.5 is used:

$$\text{Crashes prevented} = N \times CR \frac{(\text{ADT after improvement})}{(\text{ADT before improvement})} \qquad (8.5)$$

where

N   = expected number of crashes if countermeasure is not implemented and traffic volume remains the same
CR = overall crash reduction factor for multiple mutually exclusive countermeasures at a

single site

$$= CR_1 + (1\text{-} CR_1) \times CR_2 + (1\text{-} CR_1) \times (1\text{-} CR_2) \times CR_3 + \ldots + (1\text{-} CR_1) \times \ldots$$
$$\times (1\text{-} CR_{m\text{-}1}) \times CR_m = (1 - CMF) \times 100$$

CR_i = crash reduction factor for a specific site

m = number of countermeasures at the site

Therefore, the influence of countermeasures such as modifying the lane width and incorporating rumble strips into the shoulder can be incorporated into a calculation for the number of crashes prevented.

Example Problem 8.4

For a rural, two-lane, two-way highway, determine the crash reduction factor for reducing the lane width from 12 ft to 9 ft, for adding centerline rumble strips, and for the combination of these two treatments.

First, for reducing the lane width, using Table 10-8 on page 10-24 of the Highway Safety Manual, we assume that the AADT is greater than 2,000 vehicles, which provides a CMF = 1.50. The crash reduction factor is:

$$CR = (1 - CMF) \times 100 = (1 - 1.50) \times 100 = -50\%$$

Reducing the lane width from 12 ft to 9 ft increases the crashes by 50% if the ADT stays constant.

Second, for adding centerline rumble strips, on p. 10-29 the CMF = 0.94. The crash reduction factor is:

$$CR = (1 - CMF) \times 100 = (1 - 0.94) \times 100 = 6\%$$

Adding centerline rumble strips decreases the crashes by 6% if the ADT stays constant.

Finally, looking at the two treatments combined:

$$CR_T = CR_1 + (1 - CR_1) \times CR_2 = -0.50 + (1 - -0.50) \times 0.06 = -0.41 \rightarrow -41\%$$

Therefore, reducing the lane width and adding centerline rumble strips is expected to increase the crashes by 41% if the ADT stays constant.

HOMEWORK PROBLEMS

1. Asphalt Incorporated, a laboratory testing firm of asphalt materials, received two samples for testing. The mixtures were identical, except that one mixture contained 78% virgin material, 20% RAP, and 2% RAS, while the second mixture contained 100% virgin material. Please indicate which mixture you believe has the recycled material based on Superpave Performance-Graded (PG) Binder Grading testing run-off extracted asphalt binder (extraction testing following ASTM D2172) and justify your reasoning. Note – you must explain the process that you followed to grade the binder.

Extracted asphalt binder 01:

Flash Point		237°C
Viscosity at 135°C		2.84 Pa-s
Dynamic Shear G* / sin δ (unaged)	64°C	0.47 kPa
	58°C	0.94 kPa
	52°C	1.67 kPa
RTFO Mass Loss		0.97 %
Dynamic Shear G*/sin δ (RTFO aged)	64°C	2.13 kPa
	58°C	2.97 kPa
	52°C	3.31 kPa
Dynamic Shear G* x sin δ (PAV aged)	19°C	5012 kPa
	22°C	4942 kPa
	25°C	4855 kPa
Creep stiffness	-24°C	324 MPa/0.301
(S/m-value)	-18°C	298 MPa/0.305
	-12°C	253 MPa/0.322
Direct Tension	-18°C	0.87 %
(failure strain)	-12°C	1.02 %
	-6°C	1.21 %

Extracted asphalt binder 02:

Flash Point		239°C
Viscosity at 135°C		2.73 Pa-s
Dynamic Shear G* / sin δ (unaged)	64°C	0.95 kPa
	58°C	1.32 kPa
	52°C	1.87 kPa
RTFO Mass Loss		0.84 %
Dynamic Shear G* / sin δ (RTFO aged)	64°C	1.79 kPa
	58°C	2.22 kPa
	52°C	2.91 kPa
Dynamic Shear G* x sin δ (PAV aged)	19°C	5113 kPa
	22°C	5017 kPa
	25°C	4922 kPa
Creep stiffness	-24°C	312 MPa / 0.285
(S / m-value)	-18°C	295 MPa / 0.302
	-12°C	275 MPa / 0.311
Direct Tension	-18°C	0.71 %
(failure strain)	-12°C	0.92 %
	-6°C	1.04 %

2. A laboratory testing firm of asphalt materials in Madison, Wisconsin, received two samples for testing. The mixtures were identical, except that one mixture contained 65% virgin material and 35% RAP, while the second mixture contained 100% virgin material. Please indicate which mixture you believe has the recycled material based on Superpave Performance-Graded (PG) Binder Grading testing run-off extracted asphalt binder (extraction testing following ASTM D2172) and justify your reasoning. Note – you must explain the process that you followed to grade the binder.

Extracted asphalt binder 01:

Flash Point		274°C
Viscosity at 135°C		2.88 Pa-s
Dynamic Shear G* / sin δ (unaged)	64°C	0.55 kPa
	58°C	1.03 kPa
	52°C	1.57 kPa
RTFO Mass Loss		0.85 %
Dynamic Shear G* / sin δ (RTFO aged)	64°C	2.03 kPa
	58°C	2.17 kPa
	52°C	2.51 kPa
Dynamic Shear G* x sin δ (PAV aged)	13°C	5112 kPa
	16°C	4822 kPa
	19°C	4735 kPa
Creep stiffness	-24°C	325 MPa / 0.311
(S / m-value)	-18°C	288 MPa / 0.315
	-12°C	254 MPa / 0.322
Direct Tension	-24°C	0.97 %
(failure strain)	-18°C	1.03 %
	-12°C	1.11 %

Extracted asphalt binder 02:

Flash Point		242°C
Viscosity at 135°C		2.67 Pa-s
Dynamic Shear G* / sin δ (unaged)	64°C	0.85 kPa
	58°C	1.23 kPa
	52°C	1.57 kPa
RTFO Mass Loss		0.87 %
Dynamic Shear G* / sin δ (RTFO aged)	64°C	1.87 kPa
	58°C	2.31 kPa
	52°C	2.88 kPa
Dynamic Shear G* x sin δ (PAV aged)	13°C	5203 kPa
	16°C	5032 kPa
	19°C	4971 kPa

Flash Point		242°C
Creep stiffness	-24°C	312 MPa / 0.295
(S / m-value)	-18°C	302 MPa / 0.312
	-12°C	300 MPa / 0.321
Direct Tension	-24°C	0.78 %
(failure strain)	-18°C	0.98 %
	-12°C	1.08 %

3. A minor arterial road is being reconstructed through Urbana, Illinois. Traffic volume forecasts estimate that there will be 13,500 average annual daily traffic (AADT) in both directions during the first year of operation. The following vehicle mix is expected:
   - Passenger cars (2000 lbs/axle, F_{Ei} = 0.0002) = 75%
   - 3-axle single-unit trucks (12,000 lb/axle, F_{Ei} = 0.189) = 20%
   - 5-axle tandem-unit trucks (18,000 lb/axle, F_{Ei} = 0.0773) = 5%

   If the growth factor is anticipated to be 14.49 (from an annual growth rate of 8% and a 10-year design period), and the percent truck volume on the design lane is 55%, compute the total ESALs for the roadway.

4. Using Figure 8.4, calculate the approximate cost of shipping by truck and train in the United States in 2010. What would happen to the costs if 10% of the truck freight was switched to rail? Describe your process of estimating the costs, along with assumptions you had to make. Use the format provided under the sidebar in Chapter 1 "Writing a High-Quality Essay," to formulate your answer.

5. Make a graph of the stopping sight distance for all eight scenarios and describe the trends. Assume a speed of 45 mph and an upward grade of 0.7%. State other assumptions you make, and use the format provided under the sidebar in Chapter 5 "Constructing a High-Quality Graph." Choose two sets of data that you find are the most interesting and discuss your thoughts using the format provided under the sidebar in Chapter 1 "Writing a High-Quality Essay," to formulate your answer.

6. As mentioned, the assumed perception reaction time that AASHTO uses is 2.5 seconds. Assume a speed of 45 mph and an upward grade of 0.7% and find the stopping sight distance. Compare this stopping distance to the data calculated in Problem 5. Do you think that AASHTO's value for perception reaction time is appropriate? Discuss your thoughts using the format provided under the sidebar in Chapter 1 "Writing a High-Quality Essay," to formulate your answer.

7. For an undivided roadway segment (section 11.7.1 of the Highway Safety Manual), determine the crash reduction factor for increasing the lane width from 10 ft to 12 ft (Table 11-11), for increasing the shoulder width from 2 ft to 4 ft (Table 11-12), and for the combination of these two treatments.

Assume the AADT is between 400 and 2,000 vehicles, and the CMFs within these two tables can be used directly in the equations utilized in Example 8.4.

8. According to the Highway Safety Manual, the comprehensive societal crash cost of a fatal collision is $4,008,900, while a disabling injury is $216,000 (Table 7-1). Do you think that these dollar amounts are reasonable? Do you think it is appropriate to assign a dollar amount to a fatality? Discuss your thoughts using the format provided under the sidebar in Chapter 1 "Writing a High-Quality Essay," to formulate your answer.

REFERENCES

AASHTO. *Highway Safety Manual.* 1st Edition. American Association of State Highway and Transportation Officials, Washington, D.C., 2010.

Cenek, P. Kean, R., Kvatch, I., Jamieson, N. *Freight Transport Efficiency: A Comparative Study of Coastal Shipping, Rail and Road Modes.* New Zealand Transport Agency research report 497, October, 2012.

Cooper, S., Mohammad, L., Elseifi, M. Laboratory Performance of Asphalt Mixtures Containing Recycled Asphalt Shingles. *Transportation Research Record: Journal of the Transportation Research Board, No. 2445*, Transportation Research Board of the National Academies, Washington, DC, pp. 94–102, 2014.

Hansen, K., Copeland, A. Annual Asphalt Pavement Industry Survey on Recycled Materials and Warm-Mix Asphalt Usage: 2009–2012. *National Asphalt Pavement Association, Information Series 138, DTFH61-13-P-00074*, 2013.

Lutkevich, P. McLean, D., Cheung, J. FHWA Lighting Handbook. Federal Highway Administration, FHWA-SA-11-22, August, 2012.

Mitropoulos, L., Prevedouros, P. Multicriterion Sustainability Assessment in Transportation: Private Cars, Carsharing, and Transit Buses. *Transportation Research Record: Journal of the Transportation Research Board, No. 2403*, Transportation Research Board of the National Academies, Washington, DC, pp. 52–61, 2014.

Ozer, H., Al-Qadi, I., Kanaan, A., Lippert, D. Performance Characterization of Asphalt Mixtures at High Asphalt Binder Replacement with Recycled Asphalt Shingles. *Transportation Research Record: Journal of the Transportation Research Board, No. 2371*, Transportation Research Board of the National Academies, Washington, DC, pp. 105–112, 2013.

Peterson, J. A Review of the Fundamentals of Asphalt Oxidation Chemical, Physicochemical, Physical Property, and Durability Relationships. Transportation Research Circular, E-C140, Transportation Research Board of the National Academies, September 2009.

Porter, M., Irani, P., Mondor, T. Effect of Auditory Road Safety Alerts on Brake Response Times of Younger and Older Male Drivers, A Simulator Study. *Transportation Research Record: Journal of the Transportation Research Board, No. 2069*, Transportation Research Board of the National Academies, Washington, DC, pp. 41–47, 2008.

Rasul, I. Evaluation of Potential Transload Facility Locations in the Upper Peninsula (UP) of Michigan, Master's Report. Michigan Technological University, 2014.

TRB. Critical Issues in Transportation. Executive Committee of the Transportation Research Board, National Academies, 2013.

UN. Intelligent Transportation Systems (ITS) for Sustainable Mobility. United Nations Economic Commission for Europe, 2013.

West, R., Willis, J., Marasteanu, M. Improved Mix Design, Evaluation, and Materials Management Practices for Hot Mix Asphalt with High Reclaimed Asphalt Pavement Content. National Cooperative Highway Research Program, NCHRP Report 752, 2013.

Zhou, F., Li, H., Lee, R., Scullion, T., Claros, G. Recycled Asphalt Shingle Binder Characterization and Blending with Virgin Binders. *Transportation Research Record: Journal of the Transportation Research Board*, No. 2370, Transportation Research Board of the National Academies, Washington, DC, pp. 33–43, 2013.

9 Application
Construction Management

We do not inherit the Earth from our ancestors; we borrow it from our children.

Native American Proverb

Construction management is a broad term. It relates to the planning, organizing, directing, controlling, and staffing of a business enterprise. This involves significant coordination and regulation both internally within a company and externally with other companies, clients, organizations, and agencies. There are countless examples of applying economic, environmental, and social concepts to construction management. This chapter will cover four examples:

1. Materials and waste management
2. Sustainable standards
3. Community participation
4. Worker safety and roles

9.1 MATERIALS AND WASTE MANAGEMENT

It is estimated that from 1950 to 2010, the world population grew by 2.7 times, whereas the material consumption during this time grew by 3.7 times. A significant portion of this material consumption occurred as economies transitioned from an agriculture, biomass-based economy to an industrial, minerals-based economy. This increase of material consumption has increased the pressure on the environment and on the habitats of animals (Schaffartzik et al., 2014). There have been several tools developed to try and better understand how materials impact the environment. For example, the United Nations Environmental Program (UNEP) developed a metric to quantify the material footprint of consumption (UNEP, 2016). This metric is meant to represent, indirectly, the standard of living of a population. The metric is built on six categories, including services, transportation, construction, manufacturing, mining and energy, and agriculture/forestry/fisheries. The UNEP has recognized that global natural resource use has accelerated, trade in materials from one country to another has grown, and consumption is only increasing the use of raw materials. The new metric, "material footprint of consumption," indicates that in 2010 North America and Europe had a material footprint of 25 and 20 tons/capita respectively. The UNEP believes that fully industrialized nations stabilize in the 20–30 tons/capita range, as North America's material footprint was at 30 tons/capita in 2008 just before the global financial crisis. Most importantly, however, the rest of the world had a relatively small material footprint, as summarized in Table 9.1.

TABLE 9.1
Material Footprint of Consumption Around the World in 2010

Region/Country	Material footprint of consumption (tons/capita)
North America	25
Europe	20
China	14
Brazil	13
East Asia, Pacific, Latin America, Caribbean, West Asia (excluding China and Brazil)	9-10
Eastern Europe, Caucasus, and Central Asia	7.5
Africa	3.0

As seen in Table 9.1, the developing world has a significantly lower material foot-print than fully industrialized nations. However, as these regions continue to move toward full industrialization, their material footprint will only increase, thus increas-ing the burden on the global environment.

Another way to view material consumption is by quantifying the life cycle stages of materials and infrastructure. In Chapter 2, there was significant discussion about the six life cycle stages of both materials and infrastructure, and how mov-ing from Stage 5 to previous stages can be represented as reusing (Stage 5 to Stage 4), remanufacturing (Stage 5 to Stage 3), and recycling (Stage 5 to Stage 2). Using this framework, the highest positive impact for sustainability occurs with reusing. This hierarchy was first introduced in 1976 through the Resource Conservation and Recovery Act, which also began discussion on concepts such as "circular economy" and "zero waste," two common terms for waste management.

The circular economy can be clearly shown when examining material flows in the global economy. In Figure 9.1, the concept of a circular economy can be seen visu-ally by representing waste being recycled back into material processing. Figure 9.1 is a Sankey diagram, where the numbers (in units of gigatons per year) represent the size of the flow of material groups in 2005. Figure 9.1 has six material groups (bio-mass, fossil fuels, metals, waste rock, industrial minerals, and construction miner-als), eight stages that represent different steps of each material group, and the stocks. Key definitions of Figure 9.1 include:

- Domestic extraction – removal of raw material groups from the earth
- Materials processed – process of transforming raw materials into products
- Energetic use – energy required to process materials
- Waste rock – rock separated from metals during metal processing
- Material use – all usable materials during product use
- Short-lived products – disposable products

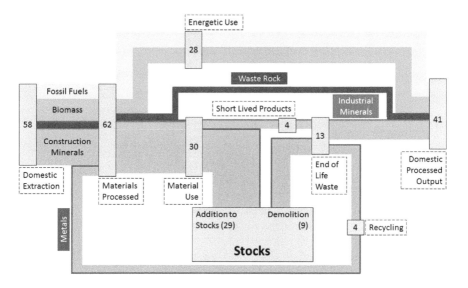

FIGURE 9.1   2005 material flows in the global economy; all numbers have units of gigatons/year (after Hass et al., 2015)

- Stocks – permanent products
  - End of life waste – products disposed at the end of their life span
  - Recycling – materials diverted from waste back to processed materials
  - Domestic processed output – material use

In Figure 9.1, it is very clear that significant strides can be made in recycling (going from end of life, Stage 5, to material processing, Stage 2), where only approximately 4 gigatons/year is recycled. However, according to Hass et al., there is no remanufacturing or reuse (going from end of life, Stage 5, to material use, Stage 3 and 4) of significance on a global scale. By implementing remanufacturing and reuse, the energetic use, waste rock, and short-lived products would all decrease by some level as well. This is a tremendous opportunity to not only increase the recycling of our material use, but ideally increase the remanufacturing and reuse as well.

In addition to quantifying the material consumption, material footprints, and material flows, there is a spectrum of material behavior. Qualification and quantification of material behavior can be loosely divided into two categories: "healthy" buildings and waste.

The first category of material behavior, healthy buildings, are buildings that are free of allergies, asthma, or other breathing problems caused by materials emitting volatile organic compounds (VOCs). Another common term for this material behavior is Indoor Environmental Quality, or IEQ. There are multiple types of materials where the VOC emissions are a concern in healthy buildings and IEQ, including adhesives, sealants, and additives to increase material performance. Table 9.2 summarizes a sampling of materials, how they are used, and limits from the South Coast

TABLE 9.2

Building Materials that Emit Undesirable VOCs

Material	Use	Allowable VOC content
Adhesives	• Bind one surface to another (floorings, wall coverings)	• <99 g/l: carpet pad, dry wall, panel, subfloor, asphalt tile, stone tile, rubber floor • 100-248 g/l: structural glazing, wood membrane, tire tread • >249 g/l: building envelop, roofing, plastic welding, rubber vulcanization, reinforced plastic
Sealants	• Fill, seal, waterproof, or weatherproof gaps and joints between two surfaces	• 250 g/l: foam insulation, grout, roadway, plumbing putty • >250 g/l: roofing, marine deck
Additives	• Added to raw materials to increase material performance	• 100 g/l: concrete curing compounds, form release compound, anti-graffiti coatings, rust prevention coating, exterior stains • >101 g/l: fire-proof coatings, sign coatings, shellac, interior stains, stone consolidates

Air Quality Management District Rules (SCAQMD) from Rule 1113 (SCAQMD, 2016) and Rule 1168 (SCAQMD, 2017).

Please take a minute to explore your surroundings. You are undoubtedly sitting on a chair of some sort that has adhesives, paints, or coatings. This chair is probably sitting on a floor. This floor could be carpeted, which means at least one of the four components of the carpet (the fiber, backing, adhesive, or pad) may contain VOCs. Alternately, the floor could be vinyl or tile, both of which require adhesives to bind to the subfloor below. And unless you are fortunate to be sitting out on a porch or in a park, you are surrounded by walls and a ceiling, which may contain paints, finishes, resins, oils, or adhesives, all of which have the potential to give off VOCs. All of these potential sources of VOCs are opportunities to use alternate materials, perhaps water-based or plant-based, that increase the healthiness and the IEQ of a building.

The second category of material behavior, waste, can be discussed in two different ways: construction and demolition (C&D) waste or hazardous waste. C&D waste accounts for approximately 30% of solid waste in the United States (Kubba, 2010), of which 85% is either recyclable or reusable material. As seen in Figure 9.1, it is estimated that 13 Gt/yr of C&D waste is generated. Research has shown that C&D waste is not necessarily as benign as once thought (Powell et al., 2015). There are more than 1,500 C&D landfills in the United States and there are no US Federal Regulations in regard to containment features, such as low-permeability liners or leachate collection systems. Powell et al. found that there were higher concentrations of dissolved solids such as sulfate chloride, iron, ammonia-nitrogen, and aluminum in the groundwater downstream from C&D landfills in Florida. These dissolved solids come from material such as wood, concrete, and gypsum drywall. However, care must be taken when considering both recycling and reusing C&D material. For example, when recycling, the point of diversion of material is important. If a material is diverted during the manufacturing process, it is called pre-consumer, or post-industrial, recycling. This material tends to have a high level of consistency. On the other hand, if a material is diverted after consumer use, it is called post-consumer recycling. The point at which a material is diverted highly influences the consistency, and potentially the quality, of the material.

For example, as discussed in Chapter 8, reclaimed asphalt shingles, or RAS, are being used in asphalt concrete mixtures (Newcomb et al., 2016). There are two types of RAS that can be used: manufacturing waste or tear-off shingles. Manufacturing waste takes the shingle material that is generated during manufacturing that is not sold, whereas tear-off shingles are taken off roofs. The shingles that have sat on a roof for up to 30 years are highly oxidized after being exposed to the sun, rain, wind, and other elements over their lifespan. In fact, it is estimated that manufactured waste has a high-temperature binder grade of approximately 130 (PG130-XX) versus the tear-off high-temperature grade of approximately 180 (PG180-XX). Therefore, it is recommended that if RAS is being used in new asphalt concrete mixtures, there are two stockpiles of roofing material in the production facility, one stockpile of manufactured waste and the second stockpile of tear-off waste. AASHTO MP 23 is the standard specification for RAS, and AASHTO PP78 provides standard practice for design considerations while using RAS. Among other information, these standards recommend that no more than 5% RAS be used in asphalt concrete mixtures by weight of aggregate. Figure 9.2 shows a picture of a stockpile of RAS, already processed for use in asphalt concrete, from the North Carolina Department of Environmental Quality (NC DEQ).

While RAS provides one example of how to divert C&D waste from the landfill and back into use in our infrastructure, there is another type of waste that requires special attention: hazardous waste. In Civil Engineering applications, there are three major types of hazardous waste: asbestos, arsenic, lead paint. Figure 9.3 shows housing with asbestos cement siding and lining.

The identification, classification, generation, and management of these three substances are fully described in 40 CFR Parts 260 – 268, known as RCRA Subtitle C regulations (Resource Conservation and Recovery Act). RCRA Subtitle C is housed under the US Environmental Protection Agency (EPA). Each of these four concepts will be discussed briefly below.

The identification and classification of hazardous waste is described in Part 261. This 250+ page portion of RCRA Subtitle C covers a lot of ground. When considering the Civil Engineering perspective, two sections stand out, the solid waste and water residue sections. The solid waste is classified as hazardous if there is more than 1 part per million in average weekly flow of spent solvents such as benzene, carbon tetrachloride, tetrachloroethylene, or trichloroethylene. A second material restriction is from methylene chloride, which must be lower than 25 parts per million. These are two examples of many that can be found in the solid waste section of Part 261. The water residues from hazardous waste can be quantified in two different

FIGURE 9.2   Reclaimed Asphalt Shingles (RAS) in North Carolina (credit: NC DEQ)

FIGURE 9.3   Asbestos cement siding and linking in housing (from wiki commons, source John M, unaltered)

ways: Toxicity characteristic leaching procedure (TCLP) or High-temperature metals recovery (HTMR). Constituents that are identified through TCLP and HTMR include antimony, arsenic, barium, beryllium, and many heavy metals such as lead, mercury, nickel, and zinc.

The generation of hazardous waste is described in Part 262. Part 262 is "only" 60 pages long, and discusses the different requirements between small-quality generators (less than 100 kg of hazardous waste a year) and large-quantity generators (more than 100 kg of hazardous waste a year). Part 262 goes on to describe transportation, recordkeeping, and reporting requirements. It finishes with preparedness, prevention, and emergency procedures for large quantity generators. Finally, the management of hazardous waste is described in Part 266. This part discusses specific hazardous wastes (many of which contain arsenic and lead paint) and how these wastes are managed. While this has been just a brief overview of hazardous waste, it is intended to provide guidance as to where more information can be found and to raise awareness of the importance of properly understanding the type of C&D waste that is generated in Civil Engineering applications.

Example Problem 9.1

In order to explore an example of materials and waste management, google the phrase "Design for Deconstruction Chartwell School." A 48-page document titled "Design for Deconstruction" should be one of the first search results. The document contains five sections, including an overview, lessons, principles and strategies, priorities, and a Chartwell School Case Study. When examining this document (Hood et al., 2015), what is the majority of lead used for, and how much is used?

Figure 1.2 shows the closed-loop material cycle for lead. While the image is, unfortunately, a bit blurry, it is obvious that the largest use of lead in this material cycle is batteries. Batteries use 1,100 thousand metric tons of lead per year.

9.2 SUSTAINABLE STANDARDS

When developing plans and drawings for a construction project, there are many concepts that need to be followed outside of the assembly or execution of the tasks in the plans and drawings. For example, ensuring that high-quality aggregate is used in Portland Cement Concrete or proper steps are taken when executing a task, there are multiple aspects of construction that need consistent guidance. This guidance often comes in the form of standards. Standards can take many forms. For example, the American Association of State Highway and Transportation Officials (AASHTO) has an extremely comprehensive document called "Standard Specifications for Transportation Materials and Methods of Sampling and Testing and Provisional Standards." Table 9.3 provides an overview of the structure of AASHTO standards.

These standards are intended to take precedence over drawings on a construction site and are considered legal documents. Therefore, the standards must be comprehensive, accurate, and clear on whichever concept is being addressed. There are dozens of standards available in the different areas of Civil Engineering. AASHTO

TABLE 9.3

An Overview of AASHTO Standards

Category	Designated letter representation	Guidance area	Example
Standard Specification	M	Critical properties of materials	AASHTO M45: "Aggregate for Masonry Mortar"
Standard Practice	R	How to execute a task	AASHTO R76: "Reducing Samples of Aggregate to Testing Size"
Standard Method of Test	T	How to run a test	AASHTO T27: "Standard Method of Test for Sieve Analysis of Fine and Coarse Aggregates"
Provisional (will eventually move to just M, R, or T)	P	Specification, practice, or method of test still in evaluation	AASHTO TP81: "Standard Method of Test for Determining Aggregate Shape Properties by Means of Digital Image Analysis"

was briefly discussed above and focuses only on transportation. In addition, agencies often have standards tailored to their needs. For example, the Arkansas Department of Transportation (ArDOT) has Standard Specifications for Highway Construction. ArDOT's standards are divided into eight divisions:

- Division 100 – General Provisions
- Division 200 – Site Preparation and Earthwork
- Division 300 – Bases and Granular Surfaces
- Division 400 – Asphalt Pavements
- Division 500 – Rigid Pavement
- Division 600 – Incidental Construction
- Division 700 – Traffic Control Facilities
- Division 800 – Structures.

These standards can be found online and can be downloaded for free by googling "ArDOT Standard Specifications for Highway Construction." On a smaller scale, the City of Fayetteville, Arkansas, also has various standards, including "Water and Sewer Standard Specifications" or "Minimum Street Standards" (both can be found online and can be download for free by googling "City of Fayetteville" plus the title of the standard). These types of agency specifications often need to be followed when performing work for public agencies, whether on a state level or local level. However, there are three well-known and extensively followed world-wide standard bodies which will be discussed in more detail below. These three bodies are ASTM

International (ASTM), the International Organization for Standardization (ISO), and the Construction Specifications Institute (CSI).

The first set of standards discussed are from ASTM International. ASTM International started in 1898 as the American Section of the International Association for Testing Materials and was renamed in 1902 as the American Society for Testing Materials (hence the ASTM acronym). However, in 2001, ASTM International was officially adopted as the name of the organization. In the late 2010s, ASTM International had over 12,500 standards, over 30,000 members, and was active in over 140 countries. Figure 9.4 shows a collection of ASTM International volumes of standards in the area of pavement materials.

While ASTM International originally began as an organization that wrote test method standards, it has since grown to include not only standards, but also proficiency testing programs (lab quality and accreditation), symposia and workshops, books and journals, and various committees. In terms of standards, ASTM International has standards and other activities in geotechnical, metals, paints and

FIGURE 9.4 Various volumes of ASTM international standards in the area of pavement materials

coatings, petroleum, plastics, rubber, textiles, and dozens of other categories. For the sake of this review, the focus will be on sustainability.

ASTM International has one primary committee on sustainability: Committee E60 – "Sustainability." However, there are various other committees that also overlap with sustainable themes, including (but not limited to):

- E27 – "Hazard Potential of Chemicals"
- E34 – "Occupational Health and Safety"
- E44 – "Solar, Geothermal and Other Alternative Energy Sources"
- E50 – "Environmental Assessment, Risk Management and Corrective Action"
- F06 – "Resilient Floor Coverings"

However, E60 is exclusively focused on sustainability. E60 has broken down sustainable specifications into six broad categories: 1) buildings and construction, 2) general sustainability, 3) hospitality, 4) sustainable manufacturing, 5) water use, and 6) conservation. ASTM International classifies over 850 of their standards as having a sustainability component, but there are only about two dozen that fall directly under E60. One specification, ASTM E3027, will be discussed as an example of how ASTM International has started focusing on sustainability.

ASTM E3027, titled "Standard Guide for Making Sustainability-Related Chemical Selection Decisions in the Life-Cycle of Products" provides guidance to product manufacturers when comparing different chemicals or ingredients across the life cycle of a product (ASTM, 2018). E3027 highlights three different methods of evaluating chemical hazards, including:

1. Clean Production Action's GreenScreen for Safer Chemicals
2. The United States Environmental Protection Agency's Design for the Environment (DtE) Alternatives Assessment Criteria for Hazards Evaluation (Safer Choice)
3. The National Academy of Sciences' A Framework to Guide Selection of Chemical Alternatives.

The standard then moves into social, economic, and ecological considerations when choosing chemicals, covering the three pillars of sustainability. In the social consideration section, social factors during the raw material acquisition stage include wages, safety and health of workers, child-labor, slave labor, worker benefits, labor practices, and other labor issues. The analysis continues for material transport, manufacturing, use, and end-of-life stages. This exercise of stepping through the life-cycle stages is repeated for the economic and environmental pillars as well. Finally, the standard finishes with a discussion on the reports that should be generated from such an analysis, including a decision report, an analysis report, and a retrospective report. While this is just one example of an ASTM International standard on sustainability, it shows how almost any civil engineering topic can be viewed through the lens of sustainability and broken down to examine the social, economic, and environmental aspects of the topic in a standard guide format that can be used by agencies and industry to better implement sustainable practices in their activities.

The second set of standards discussed are from the International Organization for Standardization, or the ISO. ISO is an organization of national standard bodies, which includes the American National Standards Institute (ANSI) in the United States. In the United States, ANSI does not actually write standards, but it accredits standard developers. Members of ANSI include the American Concrete Institute, American Society of Civil Engineering, ASTM International, and hundreds of other organizations. Like ANSI, other organizations around the world that work with standards are members of ISO. ISO was established in 1947 in order to "facilitate the international coordination and unification of industrial standards." By the late 2010s, ISO had over 22,500 standards, over 780 technical committees, and was active in over 160 countries. Like ASTM International, ISO covers all areas, from road vehicles to food products, to plastics and textiles, almost any manufactured product can be found. However, also like the review of ASTM International, the focus will be on sustainability.

There are two ISO technical committees that have a strong sustainability component: 1) ISO/TC 207 – "Environmental Management" and 2) ISO/TC 268 – "Sustainable Cities and Communities." In addition to these two technical committees, there are two standards that have strong sustainability ties as well: 1) ISO 14001 – "Environmental management systems: requirements with guidance for use" (ISO, 2015) and 2) ISO 26000 – "Guidance on social responsibility" (ISO, 2010) These two standards cover two of the three pillars of sustainability: environmental and social.

ISO 14001 is intended to aid organizations with achieving the outcomes of its environmental management system. An environmental management system consists of actions that enhance environmental performance, fulfillment of compliance obligations, and achievement of environmental objectives. These actions can be implemented in various parts of an organization, including the design, manufacture, distribution, use, and disposal of products and services. These stages of the life cycle can be viewed through emissions to air, emissions to water, the use of raw materials and natural resources, the use of energy, the generation of waste, and the use of space. While ISO14001 does not state specific environmental performance criteria, it provides a framework for an organization to implement environmentally sound decisions in an economic fashion.

ISO 26000 is quite similar in theory to ISO 14001, but instead of a focus on environmental aspects of an organization there is a focus on social aspects of an organization. ISO 26000 is built around two fundamental practices: 1) recognizing social responsibility and 2) stakeholder identification and engagement. In order to meet these two fundamental practices, ISO 26000 is tied closely to the 2015 UN Sustainable Development Metrics (often referred to as the Sustainable Development Goals, or SDG) as discussed in Chapter 1. ISO 26000 constructs a framework on social responsibility by examining six core subjects, each of which has multiple sub-issues. These are summarized in Table 9.4.

ISO 26000 then provides guidance on how to integrate these core subjects of social responsibility into an organization by exploring voluntary initiatives, communication

TABLE 9.4

ISO 26000 Six Core Subjects with Sub-issues on Social Responsibility

Core subject	Sub-issues
Human rights	• Due diligence; human rights risk situations; avoidance of complicity; resolving grievances; discrimination and vulnerable groups; civil and political rights; economic, social and cultural rights; fundamental principles and rights at work
Labor practices	• Employment and employment relationships; conditions of work and social protection; social dialogue; health and safety at work; human development and training in the workplace
The environment	• Prevention of pollution; sustainable resource use; climate change mitigation and adaptation; protection of the environment, biodiversity, and restoration of natural habitats
Fair operating practices	• Anti-corruption; responsible political involvement; fair competition; promoting social responsibility in the value chain; respect for property rights
Consumer issues	• Fair marketing, factual and unbiased information and fair contractual practices; protecting consumers' health and safety; sustainable consumption; consumer service, support, and complaint and dispute resolution; consumer data protection and privacy; access to essential services; education and awareness
Community involvement and development	• Community involvement; education and culture; employment creation and skills development; technology development and access; wealth and income creation; health; social investment

practices, actions and practices, and enhancing credibility. It is hoped that the core subjects and integration perspectives will lead to maximization of an organization's commitment to sustainable development. While ISO 26000 delves deep into each one of these areas, there is one more feature that is notable. In the appendix, there is lengthy discussion and examples of cross-sectoral indicatives. The discussion is focused on tying the six core subjects to different organizations, such as the Organization for Economic Co-operation and Development's "Risk Awareness Tool for Multinational Enterprises in Weak Governance Zones," or the United Nations Environment Program's "Life Cycle Initiative." These two organizations, along with many others, provide templates for how to explore and implement ISO's recommendations for organizations interested in integrating social responsibility throughout their own organization.

Finally, the third set of standards discussed are from the Construction Specifications Institute, or CSI. Similar to ISO, CSI works with building construction and construction material organizations to create and maintain standards that assist the construction industry's documentation and communication. CSI has developed two different formats for specification development: MasterFormat and UniFormat. These two tools are intended to aid in consistent formatting of project manuals.

Therefore, when a design firm is organizing the project details into the project manual, they know what information will be required for successful construction. Conversely, the contractors will know where to look for specific information within the project manual as they are building the infrastructure. A project manual usually has six sections:

1. General project information: contact info, table of contents, schedule of drawings, etc.
2. Bidding requirements: invitations, instructions, etc.
3. Contract forms: bonding requirements, insurance requirements, etc.
4. Contract conditions: time limits, penalties, wage rates, etc.
5. Technical specifications: materials, components, quality, etc.
6. Construction products and activities, etc.

Depending on the project, there could be more or less content than listed in these six sections, but this provides the general scope of a project manual. MasterFormat comes into play in the sixth section – construction products and activities. Obviously, there are literally thousands of construction products and activities on any given job site, so having a logical and consistent presentation of information is tremendously helpful. MasterFormat is divided into 48 divisions under five subgroups:

- General requirements (Division 01)
- Facility construction (Division 02–14) (handout, .pdf p. 22-23)
- Facility services (Division 21–28)
- Site and infrastructure (Division 29–35)
- Process equipment (Division 40–48)

Much like what was reviewed with ASTM International and ISO, it is useful for the sake of this topic to focus on sustainability, and specifically the "green" portions of MasterFormat. There are two sections of green specifications within MasterFormat. The first section, 00 31 24, is the Environmental Assessment Information. The Environmental Assessment Information section discusses soil contamination, environmental impact, and environmental impact mitigation. These sections can include information on VOC requirements, recycled content requirements, environmental installation methods, and recycling of scrap among other areas (Froeschle, 1999). The second section, 02 24 00, is the full Environmental Assessment section. This section has information on air, water, and land assessment, in addition to the chemical sampling and analysis of soils, transboundary, global environmental aspects assessment. These two sections are easy to find in any project manual that uses MasterFormat and allow for easy implementation of these sustainable concepts.

The second format for specification development from CSI is UniFormat. UniFormat is similar in theory to MasterFormat, but whereas MasterFormat focuses on construction materials, UniFormat focuses on the actual construction process (Charette and Marshall, 1999). UniFormat is divided into eight sections (CSI, 2010):

1. Substructures
2. Shell
3. Interiors
4. Services
5. Equipment and furnishings
6. Special construction and demolition
7. Site work
8. General

While the UniFormat does not have any specific sections on sustainability, there is guidance provided for sustainable design requirements under multiple sections, including (but not limited to) domestic water distribution, heating, cooling, HVAC distribution systems, and lighting. In addition, there are resources for reporting requirements leading toward sustainable design. One common way to quantify sustainable design is through green building rating systems, which will be discussed in Chapter 10. CSI also has a publication titled Sustainable Design and Construction practice guide that provides information for roles and expectations of team members, best practices, green products, project delivery, and documentation. Finally, in addition to green building rating systems and CSI's practice guide, there are dozens of sustainable practices that can be incorporated into all eight sections along the lines of topics that have been discussed in the environmental, geotechnical, structural, and transportation chapters of this textbook.

Example Problem 9.2

Along with ASTM, ISO, and CSI, Global Reporting Initiative (GRI) is another organization that has a strong focus on sustainable standards. In order to explore GRI's perspective, Google the phrase "GRI 101 Foundation 2016." A 30-page document titled "GRI 101: Foundation" should be one of the first search results. When examining this document (GRI, 2016), what are the four reporting principles for defining report content and what are the six reporting principles for defining report quality?

The four reporting principles for defining report content are: 1) stakeholder inclusiveness, 2) sustainability context, 3) materiality, and 4) completeness. The six reporting principles for defining report quality are: 1) accuracy, 2) balance, 3) clarity, 4) comparability, 5) reliability, and 6) timeliness.

9.3   COMMUNITY PARTICIPATION

Traditionally, Civil Engineering undergraduate students are well steeped in the fundamental subjects of the discipline. Environmental, geotechnical, structural, and transportational, these four areas are explored in depth in the majority of the curriculum. However, there are multiple nontechnical and educational leadership roles that engineers must play. This discussion will revolve heavily around the content in Chapter 13 (Community Participation) of "Engineering for Sustainable Communities" (Kelly et al., 2017). However, it is not limited to just these resources, as there are many other organizations that are focusing on community. For example, ASCE's five-year roadmap to sustainable development, introduced in Chapter 1, has many strains of community involvement, especially in the fourth priority "communicate and advocate."

Since many Civil Engineering infrastructure projects are large, long lasting, visible, impactful, and expensive, it is important to understand the entire lifecycle of the infrastructure project, as discussed in Chapters 2 and 3. However, if the community where the infrastructure project will be built understands and participates in the full lifecycle of the project, the involvement of the community is almost always helpful. However, communities are a dynamic system that is often challenging for traditional engineers to fully understand and leverage. Therefore, it is important to engage and use others who have expertise in political and social forces that affect infrastructure projects. Another term for these "others" are stakeholders.

Stakeholders come in many forms and are generally classified under three categories: individuals, groups, and institutions. The first category, individuals, should be treated as people, not as role players. Any individual who takes the time to come to a public hearing or community meeting often cares deeply about the topic and should be listened to and respected. The second category, groups, are individuals acting collectively. These groups can be local, regional, or across the nation. The third category, institutions, can be either public or private entities with established roles in the community. Again, institutions can be at a local, regional, or national level, and can be owners, regulators, or decision makers. Examples of institutions are boards, city committees, or other social service agencies. While there is some grey area between groups and institutions, the largest point of differentiation is that the institutions generally have established and formal roles in the community.

In order for stakeholders to be fully effective, they need to know the who/what/where/why/how of the project (Beierle and Konisky, 1999). Who will the project benefit? What is going on and what alternatives are being considered? Where will the project be performed? Why is the project being done? How will the project influence the stakeholders? These types of questions, if answered well, can fully leverage the benefits of stakeholders on a project. For example, stakeholders can help in all stages of a project. During the design phase, stakeholders can define the needs and concerns of the community and can provide an understanding of the social and political forces in a community. During the building phase, stakeholders can help prioritize and rank options and then, after feeling included in the decision-making process, can help explain choices and build support within the community. Finally, during the operation phase, stakeholders can monitor the implementation, identify changing community concerns, and assist with reevaluation, renegotiation, and repurposing of the infrastructure project.

So far the benefits of stakeholder involvement have been discussed. However, there are some challenges with stakeholders. First, many existing laws and legal framework about community participation, and the involvement of stakeholders are more than a generation old. This means that the laws and legal framework were written before internet use was widespread and before social media was established. Therefore, the majority of "formal" methods of stakeholder involvement include fixed events, such as hearings and comment periods, which are advertised by signs on the property or announced in a printed newspaper. While there has been a shift to more electronic forms of communication (Grossardt et al., 2003), like everything in Civil Engineering, the change is slow in coming. A second challenge with stakeholders is the high level of variability between the individuals, groups, and institutions. This variability comes in the form of knowledge (from highly knowledgeable to incorrect to misinformed), resources (time and money), and attitudes (constructive, obstructive, or destructive). Therefore, it is important to go into any event with stakeholders with an open mind and the determination to treat each stakeholder with a high level of respect.

Once all of the benefits and challenges associated with stakeholders are recognized, there are six factors that shape the stakeholder involvement. These factors with a brief explanation are:

1. Scale: larger projects require higher stakeholder involvement
2. Complexity: cultural/environmentally sensitive setting, level of disruption during construction, duration of the project
3. Controversy: past events may raise complex social/political issues
4. Novelty/innovativeness: new engineering techniques/unfamiliar process may raise concerns
5. Participatory culture: community history in decision making
6. Resources: budget, skill level of people interacting with public

Another way to view stakeholder involvement is through the anticipated level of involvement. For example, stakeholder involvement is generally low if the project is small, routine, and uncontroversial. However, as stakeholder involvement increases, there may be a need for public information sessions, town hall sessions, hearings, and a web-based or social media campaign. Finally for very large, high-profile, complex, and controversial projects, a high level of stakeholder involvement is necessary. For these types of projects, stakeholders should be actively sought out. There are a slew of different types of groups and institutions that should be approached, including (but not exclusively): property owners, neighborhood associations, business operators, Chambers of Commerce, or political/faith/nonprofit groups. If projects are unique enough, it may be necessary to develop special topics or advisory working groups, or even to hire consultants in order to interact with the community.

An example of a project that required a high level of stakeholder involvement was the Fayetteville Public Library expansion in Fayetteville, AR. For this project, two consultants were hired in order to obtain full stakeholder engagement. Figure 9.5 shows the existing library and the expansion under construction.

The Fayetteville Public Library is a high-profile and popular building in Fayetteville AR as it is a hub for the community. The expansion budget is estimated to be $49 million and adds a youth service division, a multipurpose area, a rooftop garden, an open-air plaza, and a "maker" space. The maker space includes filming/editing/green screens for video projects, a robotics and electronic lab, a tree house for kids, and a tool checkout. Since over $26 million of this cost was to be funded

(a)   (b)

FIGURE 9.5   The existing Fayetteville Public Library (left) and the expansion (right)

from a special tax, the expansion was also controversial. Therefore, this project checks all of the boxes (expensive, large, controversial, and high profile) for a project that warrants significant stakeholder involvement.

The Fayetteville Public Library 2030 Master Plan (which can be found online by googling the phrase) goes into significant detail on various subjects, including:

- Where Have We Been
- Measuring Success
- Facility Expansion
- Library Service Goals and Space Needs – A Planning Model
- Public and Staff Listening Sessions
- Cohort Groups for Peer Comparison

In regard to the role of stakeholders in the project, there are several pages of transcripts from the public and staff listening sessions. Table 9.5 summarizes some of the comments made during public meetings, meetings with local officials, meetings with local residents, a lunch with library volunteers, and staff responses.

As seen in Table 9.5, there is very little discussion of the geotechnical site evaluation, the hydrological design of drainage, the loading conditions for the new addition, or the transportation systems planning. The two comments found on parking were in fact contradictory: one requesting more parking, so it is easier to drive, while the second requesting less parking to induce more demand for alternate modes of transportation. Regardless, much of what is learned at the undergraduate level of Civil Engineering is not of interest to the majority of stakeholders, yet these types of meetings and sessions are essential for the successful implementation of these types of projects.

Another potential resource to quantify and qualify stakeholder participation is from a report put out by the Construction Industry Institute (CII). In brief, CII is a consortium of more than 140 leading owner, engineering-contractor, and supplier firms from both public and private arenas. CII's mission is to "Provide a research and development platform to create and drive innovative solutions that tangibly improve

TABLE 9.5

A Sampling of Stakeholder Comments on the Fayetteville Public Library

Stakeholder	Venue	Comment (may be paraphrased)
Female citizen	Public meeting	I would rather have the book shelves raised. I don't squat well any more
Male citizen	Public meeting	I have a problem with digital technology, censorship can be enacted, someone is controlling our access to information
Mayor of Fayetteville, AR	Private meeting	The biggest complaint that the Mayor's office gets from the public is the need to expand operating hours
Staff	Private meeting	The top three issues are space, study rooms, and shelving
Tweens/Teens	Private meeting	I'd like smart board for homework
Volunteers	Appreciation brunch	Expansion must include increased parking. Strong response to parking issue
Female citizen	Public meeting	I am handicapped and we get all the good parking spaces, but there are a lot of data that more parking is counterproductive; discourages public transportation, and walking, causes destruction to the earth – how are you addressing parking

business outcomes through an academically-based, disciplined approach." One way that CII executes this mission is through research reports. Research report 304-11, titled "Study of Sustainability Opportunities during Construction," is one such report (O'Connor et al., 2014). The majority of this report dealt with the development of fifty-four Construction Phase Sustainability Actions (CPSA). The project team defined construction sustainability as "processes, decisions, and actions during the construction phase of capital projects that enhance current and future environmental, social, and economic needs, while considering project safety, quality, cost, and schedule." To this end, each CPSA provided multiple sections, including:

- A primary construction function for the CPSA
- A secondary construction function for the CPSA
- A description
- A qualification of the impact on sustainability (economic, environmental, and social)
- The impact of the CPSA on safety, quality, cost, and schedule
- The ease of implementation
- Project conditions that leverage benefits from the CPSA
- Potential output metrics
- Barriers to implementation
- References.

TABLE 9.6
A Summary of CPSA 2 "Stakeholder Engagement Plan"

Section	Summary
Construction function	Primary: project management
	Secondary: none
Description	Poor stakeholder engagement leads to sustainability-related risks. Formal assessments, engagement plans, and monitoring can help.
Qualification of impact on sustainability	Economic – negligible
	Environmental – small positive (noise, light pollution, odors)
	Social – small positive (community relationships, infrastructure, traffic)
Impact	Significantly positive on project schedule performance
Ease of implementation	Challenging
Project conditions that benefit	Stakeholders clearly defined and accessible, stakeholders have diverse interests, project large and complex
Potential output metric	Percent of community issues addressed, percentage of stakeholder engagement plan implemented
Barriers	Stakeholders difficult to identify and do not wish to participate in engagement opportunities

Each CPSA was placed under one of the eight primary construction functions. The construction functions were defined as: 1) project management, 2) contracting, 3) field engineering, 4) site facilities and operations, 5) craft labor management, 6) materials management, 7) construction equipment management, and 8) quality management, commissioning, and handover. Of the 54 CPSAs developed, two were directly related to stakeholder involvement. These two were CPSA 2 (titled "Community social Responsibility Program") and CPSA 6 (titled "Stakeholder Engagement Plan"). A brief summary of CPSA 2 is found in Table 9.6.

As seen in Table 9.6, it is very apparent that involving stakeholders in planning is not easy, but it does provide sustainable benefits. Therefore, on very large, high-profile, complex, and controversial projects, such as the Fayetteville Public Library, benefits can come from establishing and implementing a stakeholder engagement plan.

Example Problem 9.3

There are literally thousands of examples of notes from public meetings that show concerns that residents have with large infrastructure projects. For example, google the phrase "Port of Seattle sustainable airport master plan may 30 open house meeting notes" and a 7-page document of the meeting notes should be on of the first search results. When examining this document (Port of Seattle, 2018), what were the two most important airport functions and the two highest priorities (including percentages) according to the submitted comment cards?

The two most important airport functions according to the submitted comment cards were: 1) environmental stewardship, 81.8% and 2) commercial airline service, 54.6%. The two highest priorities according to the submitted comment cards were: 1) aircraft noise, 92.9% and 2) air quality, 78.6%.

9.4   WORKER SAFETY AND ROLES

Worker safety is paramount on any construction site. Whether vertical or horizontal construction, it is critical that all facets of construction have a strong component of safety. There are many reasons that work safety and safety awareness has increased over the years. For example, the Workers' Compensation Laws enacted in the early 20th century transferred the responsibility for worker injury from the employee to the employer. This concept continued to move forward in the 1970s with legislation mandating that employers provide their employees with a safe work environment. A key part of this was the Occupational Health and Safety Act (OSH Act) of 1970 which established citations and fines for failure to comply with safety regulations. For the past 30 years, additional legislation has been passed for safety on construction points, including general contractors being held accountable for subcontractors and owners carrying some burden for all construction site injuries. This legislation increased liability legislation and liability insurance cost to such a high level, peaking in the 1980s, that led to a significant effort and investment into research to improve safety. Finally, in the past 20 years, a culture change has occurred in industry participants. Changes included the recognition that employee injury is not acceptable as reducing injury is a moral imperative, and that the implementation of best practice work processes significantly impact safety performance.

In addition to these legislative and cultural developments, there are clear economic and social benefits (two of the three pillars of sustainability) to implementing worker safety on the job site (Ruttenberg, 2013). Ruttenberg investigated the effect of workers completing OSHA-10, which is a 10-hour online training intended to increase awareness of safety for construction and general industry professionals put on by the Occupational Safety and Health Administration (OSHA). Based on a survey of 195 workers, Ruttenberg found that 75% of trainees carried things on ladders before the training while only 26 did after training. Similarly, before OSHA-10 training, 37% of workers checked a scaffold to verify that it was constructed properly before use, whereas 79% check after the OSHA-10 training.

While this initial research did not capture specific economic savings associated with the surveyed workers, it did estimate that reducing injuries by 2% per year could save companies and workers $336 million a year. This saving would come in two forms: direct costs and indirect costs. Direct costs include payments for hospitals, physicians, and medicine, while there can be quite a number of indirect costs due to one indirect cost causing two to three more. Indirect costs can include:

- Transportation of injured worker to medical facility
- Loss of productivity
- Disruption of schedules

- Administrative time for investigations and reports
- Training of replacement personnel
- Wages paid to the injured workers and others for time not worked
- Cost of wages for associated supervision
- Cleanup and repair
- Adverse publicity
- Damage to equipment, tools, material
- Third-party liability claims

A third form of savings also begins the transition from the economic pillar into the social pillar. The third form of savings is in quality-of-life costs. These costs are the economic value attributed to the pain and suffering that victims and their families experience as a result of injuries or illnesses. One study (Leigh et al., 1997) estimated the average cost of an injury was just over $13,200 and the average cost of an illness was just over $32,100, which included direct, indirect, and quality-of-life costs. However, pain and suffering also has non-economic considerations. Looking at the 17 UN Sustainable Develop Goals discussed in Chapters 1 and 4, pain and suffering could influence good health and well-being (Goal 3) and quality education (Goal 4), to name a few. Therefore, increasing worker health and safety can have both economic and social benefits.

OSHA is a critical organization that has enabled these economic and social benefits. OSHA is a part of the United States Department of Labor and was established in 1970 through the Occupational Safety and Health Act from Congress. The mission of OSHA is: "Assure safe and healthful working conditions for working men and women by setting and enforcing standards and by providing training, outreach, education and assistance." Generally, each state has at least one OSHA office, each of which report to one regional office. For example, OSHA has one office in Arkansas (Little Rock), and the state of Arkansas falls in Region 6, which is headquartered in Dallas, Texas. Often, the state-level offices interact with state government bodies. Again, in Arkansas, the OSHA office in Little Rock interfaces with the Arkansas Department of Labor. It is important to note that OSHA covers most private sector workers but state and local government workers are not covered by OSHA.

There are dozens of textbooks written about work safety, and literally hundreds of resources available for download at no charge on OSHA's website (www.osha.gov). One such resource available is the "OSHA Recommended Practices" (OSHA, 2016). This document has two very nice resources, a list of how to get started and recommended practices. Between these two resources, a company can either begin a safety program or enhance an existing safety program. The first resource, a list of 10 items to get your safety program started, contains straightforward steps that you can use to get started on establishing a safety culture. The 10 steps are:

1. Set safety and health as a top priority
2. Lead by example
3. Implement a reporting system
4. Provide training
5. Conduct inspections

6. Collect hazard control ideas
7. Implement hazard controls
8. Address emergencies
9. Seek input on workplace changes
10. Make improvements

OSHA is very candid in acknowledging that there is no one safety plan that covers all workplaces, but these 10 steps provide direction on general ideas that can be tailored and optimized to your conditions. For example, in Step 9, OSHA recommends that before implementing any major changes to your workplace, it is important to get feedback from the workers to identify potential safety or health issues. A very good example of this is the use of steel toe boots. Steel toe boots are a stable piece of Personal Protection Equipment (PPE) on a job site. However, it is important to recognize that if very heavy equipment is being used, the equipment can crush the steel tip. This will literally cut toes off instead of crushing toes. While neither outcome is desirable, obtaining feedback on specific situations can lead to valuable conversation and improvements, which can be discussed in Step 10.

In addition to these 10 steps, the second resource in OSHA's Recommended Practices is the most recommended practice. Table 9.7 summarizes the core elements of the safety and health program recommended practices (OSHA, 2016).

TABLE 9.7

Core Elements of the Safety and Health Program Recommended Practices

Category	Summary
Management leadership	• Demonstrate commitment to continuous improvement • Managers make safety and health a core organization value, set goals and objectives, and provide adequate resources
Worker participation	• Workers involved in all aspects and understand their roles • Obstacles to communication removed with no fear of retaliation
Hazard identification and assessment	• Identify, assess, and build procedures for workplace hazards • Inspect and reassess to identify new hazards • Identify root causes of hazards and prioritize for control
Hazard prevention and control	• Identify and select methods for eliminating, preventing, or controlling workplace hazards • First provide engineering solutions, then safe work practices, then administrative controls, finally personal protective equipment (PPE)
Education and training	• Train all workers for understanding and recognition of workplace hazards and how to implement responsibilities and solutions
Program evaluation and improvement	• Evaluate control measures for effectiveness • Establish procedures to monitor performance, implementation, and identify shortcomings
Communication and coordination	• For employers, contractors, and staffing agencies • Provide same level of safety and health protection to all employees • Communicate hazards that may be encountered

For each of the categories in Table 9.7, OSHA's Recommended Practices provides multiple action items and tips on how to accomplish each action item. For example, an action item under management leadership is to "communicate your commitment to a safety and health program." One suggestion on how to accomplish this is by establishing a written policy, signed by upper management that clearly describes the organization's commitment to safety and health. This is just one example of an action item and how to accomplish the action item, but action items and suggestions on implementation are provided for all seven categories. Other construction organizations, such as CII, also have construction-safety best practices. While many of the concepts overlap, such as management commitment and safety education, there are unique details in CIIs that OSHA does not specifically discuss at a high level, such as drug/alcohol testing and contract type. These differences in areas of emphasis reinforce the importance of examining multiple resources when evaluating an area you are not familiar with.

In addition to the Recommended Practices document, OSHA also has dozens of Pocket Guides and Wallet Cards that have been established for quick reference of a job site. These include construction, concrete manufacturing, noise, fall prevention, hazard communication, heat outreach, reporting, and warehousing. In short, these guides generally provide an overview of potential hazards and common citations (write-ups for OSHA violations), along with various other resources, associated with the topic. Two guides will be explored here: construction and concrete manufacturing.

The first guide, the construction pocket guide, begins with a brief summary of the potential hazards for workers in construction and then lists the 10 most frequent citations that OSHA handed out when inspecting job sites (OSHA, 2005). For each citation, OSHA provides a summary of the hazard and a list of solutions. After the discussion on each of the 10 most frequent citations, a checklist is provided that can help avoid hazards on the job site. Next, there is a list of publications that users can examine and programs that users can enroll in to obtain more information on specific areas of construction safety. Finally, there are several success stories provided on the potential benefits of practicing construction safety. In order to gain an understanding of the level of content for potential hazards and common citations, Table 9.8 summarizes the 6 potential hazards and 10 most common citations in construction safety as described in the construction pocket guide, with a brief discussion on the details in the trench collapse potential hazard section and scaffolding common citation section provided afterward.

The hazards associated with trench collapse are fatality and injury. OSHA states that there are dozens of deaths each year in trench collapses around the United States and hundreds of injuries. OSHA goes on to list several potential solutions, from never entering an unprotected trench, to providing an exit at least every 25 ft of trench (in the form of a ladder, stairway, or ramp), and spoils (the material dug out in order to create the trench) should not be placed within 2 ft of the trench edge. In addition, it recommends that a professional engineer design protective systems for trenches greater than 20 ft deep and has a very nice table summarizing the allowable slopes of trenches less than 20 ft deep, based on the type of soil being trenched. For

TABLE 9.8

Potential Hazards and Common Citations in the OSHA Construction Pocket Guide (OSHA, 2005)

Potential Hazards	Ten most common citations
• Falls (from heights) • Trench collapse • Scaffold collapse • Electric shock and arc flash/arc blast • Failure to use proper personal protective equipment • Repetitive motion injuries	1. Scaffolding 2. Fall protection (scope, application, definitions) 3. Excavations (general requirements) 4. Ladders 5. Head protection 6. Excavations (requirements for protective systems) 7. Hazard communication 8. Fall protection (training requirements) 9. Construction (general safety and health provisions) 10. Electrical (wiring methods, design, and protection)

example, a Type A soil (clay) should have a maximum height/depth ratio of 0.75:1.00 (or a 53° slope angle), while a Type C soil (sand) should have a maximum height/depth ratio of 1.5:1.0 (or 34° slope angle).

The second guide, the concrete manufacturing pocket guide, also begins with a brief summary of the potential hazards for workers in construction and then lists the ten most frequent citations that OSHA handed out when inspecting job sites (OSHA, 2004). Similar to the construction guide, for each citation, OSHA provides a summary of the hazard and a list of solutions. Next, there is a list of "worker safety tips" and multiple checklists to help avoid hazards. Finally, there is a list of concrete safety and health resources that users can utilize to obtain more information on specific areas of concrete manufacturing safety. The pocket guide ends with a success story from New York that showcases the potential safety and economic benefits to implementing concepts in the guide. In order to gain an understanding of the level of content for potential hazards and common citations, Table 9.9 summarizes the 6 potential hazards and 10 most common citations in concrete manufacturing as described in the concrete manufacturing pocket guide.

Led by OSHA, there have been significant advancements in worker safety in construction, which have led to economic and social benefits. Hopefully these trends will continue in the future to bring even higher levels of safety to each construction site.

Example Problem 9.4

This example problem was the last example problem written for this textbook, and was written three months into the COVID-19 pandemic in the United States. At the time of writing, OSHA had put out a document that provided guidance for workplaces. Google "OSHA 3990-03 2020" and a document titled "Guidance

TABLE 9.9

Potential Hazards and Common Citations in the OSHA Concrete Manufacturing Pocket Guide (OSHA, 2004)

Potential Hazards	Ten most common citations
• Eye, skin, and respiratory tract irritation from exposure to cement dust • Inadequate safety guards on equipment • Inadequate lockout/tagout systems on machinery • Overexertion and awkward postures • Slips, trips, and falls • Chemical burns from wet concrete	1. Hazard communication 2. Lockout/tagout 3. Confined spaces 4. Respiratory protection 5. Guarding floor and wall openings and holes 6. Electrical wiring methods 7. Noise exposure 8. Forklifts 9. Electrical systems design 10. Machine guarding

and Preparing Workplaces for COVID-19" should appear. What are the three different ways OSHA believes COVID-19 could affect the workplace?

The three different ways OSHA believes COVID-19 could affect the workplace are: 1) absenteeism, 2) change in patterns of commerce, and 3) interrupted supply/delivery.

HOMEWORK PROBLEMS

1. Within the document "Design for Deconstruction," answer the following questions about the construction industry:
   a. What percentage of material flow in the US economy (excluding food and fuel) is consumed by the construction energy?
   b. What percentage of global carbon emissions are emitted by manufacturing cement?
   c. How many tons of copper ore are needed to produce a single ton of copper
   d. In 1996, what percentage of solid waste produced in the United States is made of Construction and Demolition debris?
2. The document "Design for Deconstruction" lists 17 strategies for deconstruction. In your opinion, what are the two most important strategies? In order to answer this question, use the format provided under Sidebar 1.2 "Writing a High-Quality Essay."
3. What are the five materials with the highest CO_2 emission factor in "Design for Deconstruction," and what is each emission factor (include the unit)?
4. The document "Design for Deconstruction" lists three options for fastening interior paneling at Chartwell School. In your opinion, which is the best

option of the three? In order to answer this question, use the format provided under Sidebar 1.2 "Writing a High-Quality Essay."

5. The document "GRI 101: Foundation" lists a total of 10 reporting principles. Which two of these reporting principles, either report content or report quality, do you think are the most important? In order to answer this question, use the format provided under Sidebar 1.2 "Writing a High-Quality Essay."

6. The document "GRI 101: Foundation" lists 10 total reporting principles. The fourth principle, completeness, offers both guidance and tests which in turn ensure the report is complete. What are the three dimensions under guidance?

7. What are the seven basic processes for sustainability reporting using the GRI Standards according to the document "GRI 101: Foundation"?

8. The document "GRI 101: Foundation" lists four potential reasons for omission of a required disclosure. Which omission do you think would be most often used in Civil Engineering projects? Use two arguments to defend your choice, and use the format provided under Sidebar 1.2 "Writing a High-Quality Essay."

9. In the "Port of Seattle" meeting notes, what were the six stations available for participants, and which four organizations were represented at the meeting?

10. One of graduate's biggest surprises is the type of questions they get at public meetings. Civil Engineering students excel at designing columns and analyzing water treatment, but are not exposed as much to the impact the infrastructure will have on a community. One of the points raised in a question in the "Port of Seattle" meeting notes was that the largest impacts of the airport were being felt by minority and low-income residents, while the largest benefits are going toward Microsoft and Amazon employees. Given what you learned in Chapter 4, what two social pillar points would you bring up to answer this question that were not given in the recorded answer? Use the format provided under Sidebar 1.2 "Writing a High-Quality Essay."

11. In the "Port of Seattle" meeting notes, how may total aircraft operations were there in 2017, and what was the average per day? What is the anticipated level of operations in 2027, both over one year and the average per day?

12. In the "Port of Seattle" meeting notes, there are three questions directly related to noise mitigation. Do you think it is the Port of Seattle's responsibility to mitigate noise from the aircraft, or the homeowners? You should find information on noise mitigation from external sources. Focus on two arguments and use the format provided under Sidebar 1.2 "Writing a High-Quality Essay."

13. In the "OSHA COVID-19" document, what are the seven basic infection prevention measures provided?

14. In the "OSHA COVID-19" document, what are the five engineering controls recommended to isolate employees from work-related hazards?

15. The "OSHA COVID-19" document lists four different levels for classifying worker exposure to SARS-CoV-2. From a Civil Engineering construction application, which level would you classify for outdoor workers and which level would you classify for indoor workers? Use the format provided under Sidebar 1.2 "Writing a High-Quality Essay."

16. Since the time this book was written, there is undoubtedly much more information known about COVID-19. Using resources from OSHA, explain what you think is currently the largest challenge facing Civil Engineering in relation to COVID-19. You must find one OSHA published document to justify your challenge and write your essay around two main discussion points. Use the format provided under Sidebar 1.2 "Writing a High-Quality Essay."

REFERENCES

ASTM E3027. Standard Guide for Making Sustainability-Related Chemical Selection Decisions in the Life-Cycle of Products. *Journal of ASTM International, West Conshohocken, PA*, 2018, doi:10.1520/E3027-18A.

Beierle, T., Konisky, D. Public Participation in Environmental Planning in the Great Lakes Region. *Resources for the Future, Discussion Paper 99-50*, September, 1999.

Charette, R., Marshall, H. UniFormat II Elemental Classification for Building Specifications, Cost Estimating, and Cost Analysis. *NIST, National Institute of Standards and Technology, US Department of Commerce, NISTRI 6389*, October 1999.

CSI. *UniFormat, A Uniform Classification of Construction Systems and Assemblies.* Constructions Specification Institute, Alexandria, VA, 2010.

Froeschle, L. Environmental Assessment and Specification of Green Building Materials. *Construction Specifier*, pp. 53–57, October 1999.

GRI. GRI 101: Foundation. *Global Sustainability Standards Board, Global Reporting Initiative, GRI Standards*, 2016.

Grossardt, T., Bailey, K., Brumm, J. Structured Public Involvement, Problems and Prospects for Improvement. *Transportation Research Record: Journal of the Transportation Research Board*, 2003, 1858, 95–102.

Hass, W., Krausmann, F., Wiedenhofer, D., Heinz, M. How Circular is the Global Economy? An Assessment of Material Flows, Waste Production, and Recycling in the European Union and the World in 2005. *Journal of Industrial Ecology*, 2015,19(5), 765–777.

Hood, T., Priselac, A., van Gendt, S., Atkins, D., Melton, M., Schell, S., Gutierrez, O., Fisher, L., Guy, B., Mar, D., Mannik, H. Design for Deconstruction. United States Environmental Protection Agency, 2015.

ISO 14001. *Environmental Management Systems: Requirements with Guidance for Use.* International Organization for Standardization, Geneva, Switzerland, 3rd Edition, 2015.

ISO 26000. *Guidance on Social Responsibility.* International Organization for Standardization, Geneva, Switzerland, 1st Edition, 2010.

Kelly, W., Luke, B., Wright, R. *Engineering for Sustainable Communities – Principles and Practices.* American Society of Civil Engineers (ASCE) Press, Reston, VA, 2017.

Kubba, S. *Green Construction Project Management and Cost Oversight.* Elsevier, Burlington, MA, 2010.

Leigh, J., Markowitz, S., Fahs, M. Occupational Injury and Illness in the United States Estimates of Costs, Morbidity, and Mortality. *Archives of Internal Medicine*, 1997, 157(14), 1557–1568.

Newcomb, D., Epps, J., Zhou, F. Use of RAP & RAS in High Binder Replacement Asphalt Mixtures: A Synthesis. Special Report 213, National Asphalt Pavement Association, 2016.

O'Connor, J., Blackhurst, M., Torres, N., Woo, J. Study of Sustainability Opportunities during Construction. *Construction Industry Institute, CII Research Team 304*, CII Research Report 304-11, September, 2014.

OSHA. Worker Safety Series: Concrete Manufacturing. *Occupational Safety and Health Administration Pocket Guide, OSHA 3221–12N*, 2004.

OSHA. Worker Safety Series: Construction. *Occupational Safety and Health Administration Pocket Guide, OSHA 3252–05N*, 2005.

OSHA. Recommended Practices for Safety and Health Programs. *Occupational Safety and Health Administration, OSHA 3885*, October 2016.

Port of Seattle. Meeting Notes. *Port of Seattle, Sustainable Airport Master Plan, May 30 Open House and Presentation*, 2018.

Powell, J., Jain, P., Smith, J. Does Disposing of Construction and Demolition Debris in Unlined Landfills Impact Groundwater Quality? Evidence from 91 Landfill Sites in Florida. *Environmental Science and Technology*, 2015, 49, 9029–9036.

Ruttenberg, R. The Economic and Social Benefits of OSHA-10 Training in the Building and Construction Trades. *The Center for Construction Research and Training, CPWR*, May 2013.

Schaffartzik, A., Mayer, A., Gingrich, S., Eisenmenger, N., Loy, C., Krausmann, F. The Global Metabolic Transition: Regional Patterns And Trends of Global Material Flows, 1950–2010. *Global Environmental Change*, 2014, pp. 87–97.

SCAQMD. Architectural Coatings. *Rule 1113, South Coast Air Quality Management District, Amended February 5*, 2016.

SCAQMD. Adhesive and Sealant Applications. *Rule 1168, South Coast Air Quality Management District, Amended October 6*, 2017.

UNEP. Global Material Flows and Resource Productivity, Assessment Report for the UNEP International Resource Panel. *United Nations Environmental Program, International Resource Panel*, 2016.

10 Sustainable Certification Programs

We cannot solve our problems with the same thinking we used when we created them.

Albert Einstein

Throughout this textbook there have been countless examples and discussions of the various facets of sustainability in Civil Engineering. For example, in Chapter 5, sustainable concepts around Low Impact Development were discussed, while in Chapter 6 various types of retaining walls were explored to hold back earth fills. Chapter 7 introduced the building material cross-laminated timber and Chapter 8 tied crash modification factors to sustainability. Research papers and technical reports were the primary source of information for these topics, but there is another way to quantify the performance of infrastructure: certification programs. While there are literally dozens of sustainable certification programs available, also known as sustainable rating systems, three specific sustainable certification programs will be discussed here, along with several rating systems that lean toward the business side of companies. These sustainable certification programs can enhance the design process, reduce risk and increase accountability of implementing sustainable practices, improve the quality of life of citizens who utilize the infrastructure, and finally these sustainable certificate programs offer recognition to the owners of the infrastructure and the contractors who built the infrastructure. The three sustainability certification programs discussed will be Leadership in Energy and Environmental Design (LEED), Envision, and GreenRoads. The rating systems on the business side are a part of Environmental, Social, and Governance, or ESG.

10.1 LEADERSHIP IN ENERGY AND ENVIRONMENTAL DESIGN (LEED)

LEED has traditionally focused on certifying buildings. The first version of LEED was released in 1993 by the United States Green Building Council (USGBC). Today, the vision of LEED is "to transform the way buildings and communities are designed, built and operated, enabling an environmentally and socially responsible, healthy, and prosperous environment that improves the quality of life." This relatively broad vision has prompted LEED to offer four different types of building projects that can be certified:

1. Building Design and Construction (BD+C)

2. Interior Design and Construction (ID+C)
3. Building Operations and Maintenance (O+M)
4. Neighborhood Development (ND)

There are dozens of categories under these four types of building projects. For example, BD+C can certify new construction, major renovation, or just core and shell development. Building types under BD+C include schools, retail, data centers, and various other types of buildings. In a similar fashion, ID+C is used for commercial interiors, retail, and hospitality, while O+M includes retail, schools, hospitality, warehouses, and multi-family units. Finally, ND can certify neighborhood plans or existing built projects. The BD+C building projects are the most established LEED projects, so the following discussion will focus only on the BD+C rating system. However, the USGBC LEED website has extensive information on all four of the building project types.

The LEED system is based on points. In short, various categories have been established that quantify the sustainability of a building, and those points are added up for a final score. As LEED has been developed, some credits within the categories have moved from optional to required. Based on the final score, a rating is established. In 2020, there were four levels of LEED certification based off of 110 points (following version 4, which was released in 2013). If a building earned 80 or more points, it is considered LEED Platinum certified. If the building earned 60–79 points, it is considered LEED Gold certified. A LEED Silver certification is awarded after earning 50–59 points, and finally, 40–49 points earns LEED certified. BD+C certification can be applied to 11 different types of building constructions:

1. New construction and major renovation
2. Core and shell
3. Schools
4. Retail
5. Data centers
6. Warehouse and distribution centers
7. Hospitality
8. Healthcare
9. Homes and multifamily low-rise
10. Multifamily mid-rise
11. Transit Stations

Each of these 11 types of building construction is then broken down into eight categories:

1. Location and transportation
2. Sustainable sites
3. Water efficiency
4. Energy and atmosphere
5. Materials and resources

6. Indoor environmental quality
7. Innovation in design
8. Regional priority

Each type of building construction has slightly different requirements under the eight categories, but general concepts and themes are similar for each type of building construction. Note, the LEED 2019 reference is one of the seven guides available on usgbc.org. Table 10.1 summarizes the checklist for new construction and major renovation for BD+C (LEED, 2019), and the total number of points possible under each category and credit. User guides are also available for the other 10 types of building construction and are available on USGBC's website.

For new construction and major renovation, there are 56 credits that can be achieved. On the USGBC website, each credit has an intent and requirement list. For example, for new construction, the intent of the outdoor water use reduction, which is required, is to reduce outdoor water consumption. The requirement can be met with one of the following options: no irrigation required or reduce irrigation. For the first option, no irrigation required, the "landscape should not require a permanent irrigation system beyond a maximum two-year establishment period." For the second option, the "project's landscape water requirement must be reduced by at least 30% from the calculated baseline for the site's peak watering month." This can be achieved through plant species selection and irrigation system efficiency, and can be calculated using the EPA's WaterSense Water Budget Tool. There are additional requirements that nonvegetated surfaces (permeable or impermeable pavement) should be excluded from the landscape area calculations, and athletic fields, playgrounds, and food gardens may or may not be included.

Another credit example for new construction is the bicycle facilities. This credit is not required, but one point can be obtained if followed. The intent of the bicycle facilities is to "promote bicycling and transportation efficiency and reduce vehicle distance traveled" and to "improve public health by encouraging utilitarian and recreational physical activity." A summary of the requirements for bicycle facilities is that the project is within 200 yards (180 m) of a bicycle network and that various requirements for bicycle storage and shower rooms are met, including short- and long-term bicycle storage, location of bicycle storage, and a bicycle maintenance program. While the USGBC website is not quite the most intuitive website to navigate, googling the phrase "usgbc bicycle facilities new construction" brings the summary discussion above up as the first link. Therefore, if any of the credits in Table 10.1 are of particular interest, it is recommended to google the credit inside the phrase provided above, and a full intent and requirement should be one of the first links provided.

Another interesting feature is the ability to search LEED certified buildings within USGBC's website. By going to www.usgbc.org/projects, or by googling the name of a building, the state, and LEED, it is possible to examine the checklists of all LEED-certified buildings. Table 10.2 provides a summary of the LEED certified checklist for Hillside Auditorium at the University of Arkansas.

TABLE 10.1

Summary of LEED New Construction and Major Renovation Credits

Category (110 Points Total)	Credit	Points Possible
No category (1 point total)	Integrative process	1
Location and transportation (16 points total)[a]	Sensitive land protection	1
	High-priority site	2
	Surrounding density and diverse uses	5
	Access to quality transit	5
	Bicycle facilities	1
	Reduced parking footprint	1
	Green vehicles	1
Sustainable sites (10 points total)	Construction activity pollution prevention	Required
	Site assessment	1
	Site development—protect or restore habitat	2
	Open space	1
	Rainwater management	3
	Heat island reduction	2
	Light pollution reduction	1
Water efficiency (11 points total)	Outdoor water use reduction, indoor water use reduction, building-level water metering	Required
	Outdoor water use reduction	2
	Indoor water use reduction	6
	Cooling tower water use	2
	Water metering	1
Energy and atmosphere (33 points total)	Fundamental commissioning and verification, minimum energy performance, building-level energy metering, fundamental refrigerant management	Required
	Enhanced commissioning	6
	Optimized energy performance	18
	Advanced energy metering	1
	Demand response	2
	Renewable energy production	3
	Enhanced refrigerant management	1
	Green power and carbon offsets	2
Materials and resources (13 points total)	Storage and collection of recyclables, construction and demolition waste management planning	Required
	Building life-cycle impact reduction	5
	Building product disclosure and optimization – environmental product declarations	2
	Building product disclosure and optimization – sourcing of raw materials	2
	Building product disclosure and optimization – material ingredients	2
	Construction and demolition waste management	2

(Continued)

TABLE 10.1 (CONTINUED)
Summary of LEED New Construction and Major Renovation Credits

Category (110 Points Total)	Credit	Points Possible
Indoor environmental quality (16 points total)	Minimum indoor air quality performance, environmental tobacco smoke control	Required
	Enhanced indoor air quality strategies	2
	Low-emitting materials	3
	Construction indoor air quality management plan	1
	Indoor air quality assessment	2
	Thermal comfort	1
	Interior lighting	2
	Daylight	3
	Quality views	1
	Acoustic performance	1
Innovation (6 points total)	Innovation	5
	LEED accredited professional	1
Regional priority (4 points total)	Regional priority: specific credit	4

[a] A project applying for a LEED Neighborhood Development Location applies for an umbrella 16 points and does not need to pursue each subcredit.

Hillside Auditorium received 53/110 points, which places it as LEED Silver certified. Another example of a LEED certified building is the Nanoscale Science Engineering Building, which was awarded certification in May 2012. The Nanoscale Science Engineering Building is LEED Gold certified, receiving 42/69 points at the time of certification. This certification utilized a previous LEED rating system, which had a different point value than Hillside Auditorium. These two buildings are shown in Figure 10.1.

TABLE 10.2
Summary of LEED Certified Checklist for Hillside Auditorium

Category	Points (53/110)
Sustainable sites	18/26
Water efficiency	6/10
Energy and atmosphere	12/35
Materials and resources	6/14
Indoor environmental quality	8/15
Innovation	1/6
Regional priority credits	2/4
Integrative process credits	0/3

(a)   (b)

FIGURE 10.1   Hillside Auditorium (LEED Silver) (a) and the Nanoscale Science Engineering Building (LEED Gold) (b) in the University of Arkansas campus. (Credit: A. Braham.)

Example Problem 10.1

In order to gain a perspective the economic impact of green buildings, google the phrase "The Costs and Financial Benefits of Green Buildings." While this is an older report (Kats et al., 2003), it is considered one of the seminal documents when discussing the economics behind green buildings. When examining this document, list the three obstacles to sustainable buildings according to the Sustainable Building Task Force *Blueprint*.

The three obstacles listed are: (1) incomplete integration within and between projects, (2) lack of life cycle costing, and (3) insufficient technical information.

10.2   ENVISION

While the LEED system has been utilized in over 100,000 projects (as of November 2019), there is another certification that has gained traction in civil engineering: Envision. According to ASCE, Envision was founded in 2010 by ASCE, the American Council of Engineering Companies, and the American Public Works Association. Envision is currently administered by the Institute for Sustainable Infrastructure. Similar to LEED, the Envision system is based on points (Envision, 2018). Envision v3 has four levels of certification based off of 1000 total points. If a project earned greater than 50% of the points, it is considered Envision Platinum certified. If the project earned 40%–50% of the points, it is considered Envision Gold certified, if 30%–40% of the points were earned, it was considered an Envision Silver certification, and finally, if 20%–30% of the points were earned, it was considered Envision Verified certified. Certification can be applied to six different groups of infrastructure. The groups of infrastructure, with some specific examples, are:

1. Energy: distribution, hydroelectric, coal, natural gas, wind, solar, biomass
2. Water: treatment, distribution, capture/storage, stormwater, flood control, nutrient management
3. Waste: solid waste, recycling, hazardous, waste collection & transfer

4. Transportation: airports, roads/highways, bikes/pedestrians, railways, transit, ports, waterways
5. Landscape: public realm, parks, ecosystem services, natural infrastructure, environmental remediation
6. Information: telecom, cables, internet, phones, data centers, sensors

Each of these six groups of infrastructure is broken down into five categories. Each category also has multiple subcategories. Under each subcategory are several credits that are available. The categories, subcategories, and credits available under each category are as follows:

1. Quality of life: well-being, mobility, community – 13 credits
2. Leadership: collaboration, planning, economy – 11 credits
3. Resource allocation: materials, energy, water – 13 credits
4. Natural world: siting, conservation, ecology – 13 credits
5. Climate and resilience: emissions, resilience– 9 credits

Finally, each credit has up to five levels of achievement, with increasing point value for each level of achievement. All five levels of achievement are not applicable to all credits. The five levels of achievement are

1. Improved (1–11 points): performance above conventional, slightly exceeds regulatory requirements
2. Enhanced (2–18 points): indications that superior performance is within reach
3. Superior (4–24 points): sustainable performance at a high level
4. Conserving (6–26 points): performance with essentially zero negative impact
5. Restorative (8–26 points): performance restores natural or social systems

Table 10.3 summarizes the checklist for Envision, and the total number of points possible under each category and credit.

Similar to LEED, each credit has a summary of characteristics that includes the intent, definitions of the level of achievement, a description, a discussion on advancing to higher achievement levels, and evaluation criteria and documentation. For example, the intent of the credit "preserve views and local character" is to "design the project to maintain the local character of the community and to not negatively impact community views." The levels of achievement are as follows:

• Improved: understanding and balance
• Enhanced: alignment with community values
• Superior: community preservation and enhancement
• Conserving: community connections and collaboration
• Restorative: restoration of community and character

TABLE 10.3
Summary of Envision Credits

Category (maximum points)	Subcategory (maximum points)	Credit (with maximum level of achievement: C = conserving, R = Restorative)	Maximum Points
Quality of Life (200)	Well-being (92)	Improve Community Quality of Life (R)	26
		Enhance Public Health & Safety (R)	20
		Improve Construction Safety (C)	14
		Minimize Noise & Vibration (R)	12
		Minimize Light Pollution (R)	12
		Minimize Construction Impacts (C)	8
	Mobility (44)	Improve Community Mobility (R)	14
		Encourage Sustainable Transportation (R)	16
		Improve Access and Wayfinding (C)	14
	Community (64)	Advance Equity & Social Justice (R)	18
		Preserve Historic & Cultural Resources (R)	18
		Enhance Views & Local Character (R)	14
		Enhance Public Space & Amenities (R)	14
Leadership (182)	Collaboration (72)	Provide Effective Leadership & Commitment (C)	18
		Foster Collaboration & Teamwork (C)	18
		Provide for Stakeholder Involvement (R)	18
		Pursue By-product Synergies (R)	18
	Planning (60)	Establish a Sustainability Management Plan (C)	18
		Plan for Sustainable Communities (R)	16
		Plan for Long-Term Monitoring & Maintenance (C)	12
		Plan for End-of-Life (C)	14
	Economy (50)	Stimulate Economic Prosperity & Development (C)	20
		Develop Local Skills & Capabilities (R)	16
		Conduct a Life-Cycle Economic Evaluation (R)	14
Resource Allocation (196)	Materials (66)	Support Sustainable Procurement Practices (C)	12
		Use Recycled Materials (C)	16
		Reduce Operational Waste (C)	14
		Reduce Construction Waste (C)	16

(Continued)

TABLE 10.3 (CONTINUED)
Summary of Envision Credits

Category (maximum points)	Subcategory (maximum points)	Credit (with maximum level of achievement: C = conserving, R = Restorative)	Maximum Points
		Balance Earthwork On Site (C)	8
	Energy (76)	Reduce Operational Energy Consumption (C)	26
		Reduce Construction Energy Consumption (C)	12
		Use Renewable Energy (R)	24
		Commission & Monitor Energy Systems (C)	14
	Water (54)	Preserve Water Resources (R)	12
		Reduce Operational Water Consumption (R)	22
		Reduce Construction Water Consumption (C)	8
		Monitor Water Systems (C)	12
Natural World (232)	Siting (82)	Preserve Sites of High Ecological Value (R)	22
		Provide Wetland & Surface Water Buffers (R)	20
		Preserve Prime Farmland (R)	16
		Preserve Undeveloped Land (R)	24
	Conservation (78)	Reclaim Brownfields (R)	22
		Manage Stormwater (R)	24
		Reduce Pesticide & Fertilizer Impacts (R)	12
		Protect Surface & Groundwater Quality (R)	20
	Ecology (72)	Enhance Functional Habitats (R)	18
		Enhance Wetland & Surface Water Functions (R)	20
		Maintain Floodplain Functions (R)	14
		Control Invasive Species (R)	12
		Protect Soil Health (R)	8
Climate and Resilience (190)	Emissions (64)	Reduce Net Embodied Carbon (C)	20
		Reduce Greenhouse Gas Emissions (R)	26
		Reduce Air Pollutant Emissions (R)	18

(*Continued*)

TABLE 10.3 (CONTINUED)
Summary of Envision Credits

Category (maximum points)	Subcategory (maximum points)	Credit (with maximum level of achievement: C = conserving, R = Restorative)	Maximum Points
	Resilience (126)	Avoid Unsuitable Development (R)	16
		Assess Climate Change Vulnerability (C)	20
		Evaluate Risk and Resilience (C)	26
		Establish Resilience Goals and Strategies (C)	20
		Maximize Resilience (C)	26
		Improve Infrastructure Integration (R)	18

The description of each credit includes a discussion about the project design and ensures the context of the credit is clear. An urban setting example of context would be the inclusion of traditional streetscapes, building material choices, or height limitations. A rural example would include discussion about views and vistas of natural landscapes, along with other prominent natural features. The main concept behind advancing to higher achievement levels revolves around the concept of simply minimizing impacts to preservation and restoration, toward a more comprehensive planning process that takes stakeholder input into account. Finally, evaluation criteria and document would include plans and drawings, specific documents that emphasize specific contextual features, and a summary of existing policies and regulations. While this is just one example of one credit, a comprehensive summary has been put together by the Institute for Sustainable Infrastructure (Envision, 2018).

A good example of the Envision certification process can be found in Eugene, Oregon, along the Alder Street Active Transportation Corridor (Rodrigues, 2013). Alder Street has a high level of pedestrian and bicycle traffic. In order to maximize safety and promote travel by these two modes, the City of Eugene partnered with the University of Oregon and Lane Transit District to completely reconstruct the corridor. By incorporating new bicycle features (two-way buffered and contraflow bicycle lanes, colored pavement), new pedestrian features (widened sidewalks, comprehensive tree canopy), and improved signalization, the city executed their goal of not only providing a safe space to travel, but also enhancing the facilities to encourage nonmotorized modes of transportation. The self-assessment provided a 65% "yes" rating for the quality of life, a 74% "yes" rating for leadership, a 34% "yes" rating for resource allocation, an 11% "yes" rating for natural world, and a 45% "yes" rating for climate. Taken together, the assessment provided an overall 59.8% "yes" rating, which would place the project at an Envision Platinum certification. Recall, however, that this was simply the self-assessment and not the certified assessment. To continue the example above, under the credit "preserve views and local character," the city scored a 2/2 by answering yes to the two assessment questions. The commentary stated that "the project team worked closely with stakeholders to replace streetlights

(a)   (b)

FIGURE 10.2   Before (a) and after (b) of the Alder Street Active Transportation Corridor. (Credit: Rodrigues, 2013)

and construct sidewalk and tree plantings that preserved and enhanced the local character." This commentary showed specific examples of how the team achieved the credit. Figure 10.2 shows a before and after image of the corridor.

Since LEED was developed approximately 20 years before Envision, there are far more LEED projects across the world versus Envision projects. However, there are similarities and differences between the two certification systems. There are several areas of overlap in application areas, which fall under credits for both systems, such as light pollution, stormwater runoff quality, and alternative transportation. In fact, when looking at the overall content, Envision essentially covers all of LEED through the resource allocation, natural world, and climate categories. The interesting comparison comes from the differences from the two systems. Overall, LEED is a binary system, where questions are either "yes" or "no." However, Envision has the five levels of achievement, which allows for not only more of a spectrum of ratings to be established, but also allows owners and agencies to strive for incremental improvements instead of simply getting an all or nothing. When filling out the checklist, LEED is more straightforward on a smaller scope; whereas, Envision is more subjective on the level of each category but provides a larger and more flexible scope.

Another difference is that taken as a whole, LEED tends to redevelop sites and materials, while Envision tends to focus on preserving resources. This may stem from the fact that LEED focuses exclusively on buildings, whereas Envision strives to include multiple different types of infrastructure (including energy, water, waste, transportation, etc.) This brings up the last point that LEED is focused primarily on the environmental pillar of sustainability, while Envision covers both the environmental and social pillar. While there are pros and cons to each certification system, it is interesting to compare them directly and will be even more interesting to see how they will evolve over time.

Example Problem 10.2

The Water Environment Federation developed a concise fact sheet tying Envision to water infrastructure; google the phrase "Use of the Envision Sustainable Infrastructure

Rating System for Water Infrastructure." When examining this document (Bowles et al., 2019), list the 11 NYCDEP Sustainability Focused Standard Operating Procedures.

The 11 operating procedures are: (1) Sustainability Management Plan, (2) Sustainability Workshop, Preliminary (Facility Planning Stage), (3) Sustainability Workshop, Deep Dive (Design Stage), (4) Sustainability Rating Systems, (5) Energy Conservation & GHG Reduction Plan, (6) Climate Risk Assessment and Adaptation Plan, (7) Recycled Content for New Materials, (8) Construction Waste Estimate Reporting, (9) GHG Emissions Calculating Tool (LGOP Based), (10) Climate Risk Mitigation Guidance, and (11) Rooftop Solar Design Guidance.

10.3   GREENROADS

While LEED focuses on structures and Envision focuses on infrastructure, GreenRoads focuses on transportation projects (Greenroads, 2019). Note, it is possible to download free of cost an abridged version of the Greenroads user manual (2019) or you can download an entire older version of the manual (2011); both can be found at Greenroads.org. Greenroads was established in 2010 and initially focused just on roads, but has since expanded (in version 2, or v2.0) to include a wide range of projects including:

- Street improvements
- Highway improvements
- Stormwater utility improvements (either in a street or within the right-of-way)
- Bridge replacements (including structural and approach work)
- Transit-related improvements and stop facilities

Both the street and highway projects include new construction, reconstruction, and rehabilitation, while the street projects also include streetscapes. Much like LEED and Envision, there are certification thresholds of which Greenroads has four: Evergreen (≥80 points), Gold (≥60 points), Silver (≥50 points), and Bronze (≥40 points). These thresholds are out of a total 130 possible points. The 130 possible points fall within 7 categories that have a blend of mandatory or voluntary credits. There are 12 mandatory credits and 49 "other" credits that include core credits and extra credits. The categories, along with credits and associated point values, are summarized in Table 10.4.

There is extensive documentation on each of these credits, but for the sake of brevity just a few will be discussed here. For example, the Quality Control credit (which falls under the Project Requirement category) is designed to encourage quality management practices during construction. In order to achieve this, the prime contractor is required to develop, implement, and maintain a Quality Control Plan (QCP). The QCP must be in place and approved by the owner before construction can begin. The QCP must, at a minimum, contain the following:

- Key quality personnel, including responsibilities and qualifications
- Procedures used to control quality (i.e. construction items to be monitored, tests to be performed)
- Corrective action plan (i.e. procedures to implement corrective action, and if necessary, procedures to modify QCP)

TABLE 10.4

Summary of Greenroads Credits

Category (maximum points)	Credit	Maximum Points
Project Requirements (n/a)	Ecological Impact Analysis	Required
	Energy & Carbon Footprint	Required
	Low Impact Development	Required
	Social Impact Analysis	Required
	Community Engagement	Required
	Lifecycle Cost Analysis	Required
	Quality Control	Required
	Pollution Prevention	Required
	Waste Management	Required
	Noise & Glare Control	Required
	Utility Conflict Analysis	Required
	Asset Management Systems	Required
Environment and Water (30)	Preferred Alignment	1–3
	Ecological Connectivity	1–3
	Habitat Conservation	1–3
	Land Use Enhancements	1–3
	Vegetation Quality	1–3
	Soil Management	1–3
	Water Conservation	1–3
	Runoff Flow Control	1–3
	Enhanced Treatment: Metals	1–3
	Oil and Contaminant Treatment	1–3
Construction Activities (20)	Environmental Excellence	1–3
	Workzone Health & Safety	1–2
	Quality Process	1–3
	Equipment Fuel Efficiency	1
	Workzone Air Emissions	1
	Workzone Water Use	1–3
	Accelerated Construction	1–2
	Procurement Integrity	1
	Communication and Outreach	1
	Fair and Skilled Labor	1–2
	Local Economic Development	1
Materials and Design (24)	Preservation and Reuse	1–5
	Recycled and Recovered Content	1–5

(Continued)

TABLE 10.4 (CONTINUED)
Summary of Greenroads Credits

Category (maximum points)	Credit	Maximum Points
	Environmental Product Declaration	2
	Health Product Declaration	2
	Local Materials	1–5
	Long-Life Design	1–5
Utilities and Controls (20)	Utility Upgrades	1–2
	Maintenance and Emergency Access	1
	Electric Vehicle Infrastructure	1–3
	Energy Efficiency	1–3
	Alternative Energy	1–3
	Lighting and Controls	1–3
	Traffic Emission Reduction	1–3
	Travel Time Reduction	1–2
Access and Livability (21)	Safety Audit	1–2
	Safety Enhancements	1–2
	Multimodal Connectivity	1–2
	Equity and Accessibility	1–2
	Active Transportation	1–2
	Health Impact Analysis	2
	Noise and Glare Reduction	1–3
	Culture and Recreation	1–2
	Archaeology and History	1–2
	Scenery and Aesthetics	1–2
Creativity and Effort (15)	Educated Team	1–2
	Innovative Ideas	1–5
	Enhanced Performance	1–5
	Local Values	1–3

Another credit example, Ecological Connectivity (which falls under the Environment and Water category), is intended to reduce accidents between vehicles and wildlife (Greenroads, 2011). This is achieved by either only rehabilitating existing infrastructure (one point) or by rehabilitating existing infrastructure while building new infrastructure (three points). Infrastructure includes culverts, fencing, or other crossing structures for wildlife.

In addition to over 100 Greenroad certified projects across the United States, there has been extensive research also performed utilizing Greenroads as a tool to quantify sustainability. For example, Pittenger (2011) applied the Greenroads standard to pavement maintenance treatments at multiple airports. In this work, nine airports from various locations, climates, and aircraft movement (volume and types) were examined. The airports were in Billings, Oklahoma City, Boston, Dallas, Orlando, Salt Lake, San Francisco, Seattle, and Toronto. This study utilized v1.0 of GreenRoads and examined how each category was applicable to the various pavement maintenance treatments. The pavement maintenance treatments used at these nine airports included micro surfacing, asphalt concrete overlays (1.5 inches or 38 mm), slurry seal, fog seal, and shotblasting. Table 10.5 provides a summary of the categories within Greenroads that were applicable to these pavement maintenance treatments at airports as well as those which were not applicable.

Table 10.5 is interesting for two reasons. First, by comparing the categories in Table 10.5 from the 2011 study to the current categories in Table 10.4 of v2.0, it is easy to see how the concept of sustainability as it relates to roads and certification has changed in the nine-year time frame. Completely new categories have been created, including a design component (within materials and design), utilities and controls, livability (within access and livability), and creativity and effort. In addition to changes within project categories, there have been developments within each category as well. For example, within the project requirements, credits have been developed for energy and carbon footprint, social impact analysis, community engagement, glare (as a part of noise and glare control), utility conflict analysis, and asset management systems. A second interesting takeaway from Table 10.5 is that, according to Pittenger's analysis, the airports could obtain up to 60 total credits, which would allow airports to obtain up to a "gold" certification level. In addition to gaining the points, however, the lone unmet project requirement (Noise Mitigation Plan) would also have to be addressed. However, since this research was only focusing on pavement maintenance treatments, it is likely an airport could pull in other dimensions of their activities in order to achieve this required credit. Overall, viewing Greenroads through the lens of airport pavement maintenance provides insights on the structure of Greenroads and how to apply the certification to pavements outside of the traditional realm of passenger cars and trucks.

A second study examined existing roads to see how "sustainable" projects differed from "conventional" projects when run through the Greenroads rating system process (Anderson and Muench, 2013). This research, like Pittenger's, used Greenroads v1.0 as shown in Table 10.5. In short, this study looked at 105 existing roadway and bridge projects to see how the projects would be rated using Greenroads certification. In order to discuss how these projects fit into the Greenroads certification, the percentage achieved on these projects will be divided into three categories: greater than 80% achieved, 20–80% achieved, and less than 20%. These three groupings will show which credits are relatively easy to achieve with standard project planning (>80% of projects achieve), which credits are being achieved on some projects but not on a large percentage (20–80%), and which credits are difficult to achieve (<20%). Table 10.6 summarizes the credit distribution within these three categories.

When examining Table 10.6, it is disheartening to see only three credits where more than 80% of the 105 projects achieved the Greenroads requirement. However,

TABLE 10.5

Applicable Pavement Maintenance Treatments at Airports using Greenroads v1.0

Greenroads Category	Applicable categories	Nonapplicable categories
Project Requirements	• Environmental Review Process • Lifecycle Cost Analysis • Lifecycle Inventory • Quality Control Plan • Waste Management Plan • Pollution Prevention Plan • Low Impact Development • Pavement Management System • Site Maintenance Plan • Educational Outreach	• Noise Mitigation Plan
Environment and Water	• Environmental Management System • Ecological Connectivity	• Runoff Flow Control • Runoff Quality • Stormwater Cost Analysis • Site Vegetation • Habitat Restoration • Light Pollution
Access and Equity	• Safety Audit	• Intelligent Transportation System • Context Sensitive Solutions • Traffic Emissions Reduction • Pedestrian Access • Bicycle Access • Transit & HOV Access • Scenic Views • Cultural Outreach
Construction Activities	• Quality Management System • Environmental Training • Site Recycling Plan • Fossil Fuel Reduction • Equipment Emission Reduction • Paving Emission Reduction • Water Use Tracking • Contractor Warranty	• n/a
Materials and Resources	• Lifecycle Assessment • Pavement Reuse • Recycled Materials • Regional materials	• Earthwork Balance • Energy Efficiency
Pavement Technologies	• Permeable Pavement • Warm Mix Asphalt • Cool Pavement • Pavement Performance Tracking	• Long Life Pavement • Quiet Pavement
Custom Credit	• Case by Case	

TABLE 10.6
Greenroads Credits Achieved in 105 Projects

Less than 20% Achieved	20–80% Achieved	Greater than 80% Achieved
• Life-Cycle Inventory • Waste Management Plan • Environmental Management System • Stormwater Cost Analysis • Habitat Restoration • Light Pollution • Safety Audit • Quality Management System • Site Recycling Plan • Fossil Fuel Reduction • Equipment Emission Reduction • Water Use Tracking • Contractor Warranty • Life-Cycle Assessment • Earthwork Balance • Energy Efficiency • Permeable Pavement • Warm Mix Asphalt • Quiet Pavement • Pavement Performance Tracking	• Environmental Review Process • Life-Cycle Cost Analysis • Quality Control Plan • Noise Mitigation Plan • Low-Impact Development • Site Maintenance Plan • Educational Outreach • Runoff Flow Control • Runoff Quality • Site Vegetation • Ecological Connectivity • Intelligent Transportation Systems • Traffic Emission Reduction • Pedestrian Access • Bicycle Access • Transit and HOV Access • Scenic Views • Cultural Outreach • Environmental Training • Paving Emission Reduction • Pavement Reuse • Recycled Materials • Regional Materials • Long-Life Pavement • Cool Pavement	• Pollution Prevention Plan • Pavement Management System • Context-Sensitive Solutions

it is exciting to see there are over 25 credits where some standard projects already are achieving Greenroads requirements. This indicates the credits within the 20–80% range may not be achieved on a wide-spread basis, but they are being achieved on enough projects where there is a high probability of being able to implement on other projects. Finally, some of the credits that were less than 20% in 2013 (i.e. energy efficiency and warm mix asphalt) have seen significant activity since 2013 and are much more widespread today. Therefore, while this study is starting to show its age a bit, it does provide clear areas of improvement on our roadway projects to increase their potential for success in Greenroads.

A third research project examined how projects were approaching sustainability by examining which credits were being leveraged in order to obtain Greenroads certification (Lew *et al.*, 2016). This work, which used Greenroads v1.5, found that by far, the two most common credits obtained on 28 projects were context sensitivity solutions and regional materials. There was a substantial drop off to the next set of credits, some of which were site vegetation, long-life pavement, cool pavement, and energy efficiency. Also of interest were 6 credits that across the 28 projects were not ever obtained:

- Habitat restoration
- Traffic emission reduction
- Fossil fuel reduction
- Equipment emission reduction
- Quiet pavement
- Pavement performance tracking

In hindsight, this list of six credits is interesting because it shows some of the thought processes behind Greenroads v2.0, which no longer provide credit for fossil fuel reduction, equipment emission reduction, quiet pavement, and pavement performance tracking. While there is discussion on these credits embedded in some of the newly developed credits, these credits are no longer explicitly required.

Overall, these research papers show the potential and power of Greenroads as a certification program for roadway infrastructure. It is hoped that, over time, agencies and owners of transportation will begin leveraging Greenroads as much as the structural community has embraced LEED certification.

Example Problem 10.3

The University of Washington produced a final report on Greenroads when it was first developed; google the phrase "Greenroads: A sustainability performance metric for roadway design and construction." When examining this document (Muench and Anderson, 2009), there is a section that discusses interoperability with other systems, where the other systems are sustainability-related initiatives, coalitions, rating systems, and procedures. In this section, nine systems are identified. What are the nine systems?

The nine systems are: (1) complete streets, (2) sustainable site initiatives, (3) Civil Engineering Environmental Quality Assessment and Award Scheme (CEEQUAL), (4) Low Impact Development (LID), (5) Context Sensitive Solutions (CSS), (6) smart growth, (7) Resource Conservation Challenge (RCC), (8) eco-logical, and (9) LEED for neighborhood development.

10.4   ENVIRONMENTAL, SOCIAL, AND GOVERNANCE (ESG)

LEED, Envision, and Greenroads are examples of certifications that can be awarded to specific Civil Engineering projects. There are LEED gold–certified data centers, Envision silver–certified water distribution systems, and Greenroads evergreen–certified highways. However, there is another perspective that can be examined in Civil Engineering: the business perspective. This discussion will focus on the concept of

Environmental, Social, and Governance (ESG, as introduced in Chapter 4) and will provide an overview of three providers that quantify ESG.

ESG has foundations in responsibly investing in companies. Instead of only looking at the financial performance of a company, investors also look at the values of a company. Some examples of values (Kocmanová and Šimberová, 2014) are:

- Environmental: investments, emissions, resource consumption, waste
- Social: society, human rights, labor practices and decent work, product responsibility
- Governance: monitoring and reporting, corporate governance effectiveness, corporate governance structure, compliance

Kocmanová and Šimberová provided performance indicators that reflected each value, in addition to providing a measurement in order to quantify each of the performance indicators. For example, the performance indicators and measurements for the environmental measurement area (E) are shown in Table 10.7. A similar table is provided for social (S) and corporate governance (G).

TABLE 10.7

Values, Indicators, and Measurements for the Environment (Kocmanová and Šimberová, 2014)

Value	Performance indicator	Measurement
Investment	• Acquired investments for environmental protection • Environmental non-investment expenditures	• Total investments by sales • Total non-investment expenditures by sales
Emissions	• Total annual emissions • Total annual emission of greenhouse gases	• Total emissions (PM, SO_2, NO_x, NH_3) divided by sales • Total direct and indirect emissions (CO_2, CH_4, N_2O, HFC, PFC, SF_6) divided by sales
Resource consumption	• Energy use • Renewable energy use • Material use • Recycled material use • Water use	• Total direct and indirect energy consumption divided by sales • Total renewable energy sources divided by total energy sources • Total consumption of materials divided by total operation costs • Percent content of used recycled materials of total consumption materials • Total water use divided by sales
Waste	• Production of waste • Production of hazardous waste	• Total waste use divided by sales • Total hazardous waste use divided by sales

It is interesting to see how the values in Table 10.7 can be mapped to the various sustainability initiatives developed by the United Nations. For example, looking at the UN Sustainability Development Goals (UN, 2015), resource consumption can be related to Goal 12 (responsible consumption and production) while human rights, under the social umbrella, can be related to Goal 3 (good health and well-being) and Goal 10 (reduced inequalities). The question may be asked why this is important. The Principles for Responsible Investment, or PRI, teamed with the United Nations (PRI, 2019) to identify three forces that are motivating the importance of responsible investment: materiality, client demand, and regulation. The first force, materiality, acknowledges that ESG factors can influence investor returns. For example, the Deepwater Horizon oil spill required BP to record over a $50 billion pre-tax charge in 2010 and Volkswagen had to pay over $30 billion in penalties and fines in 2015 because they deceived emissions tests so their diesel vehicles passed the tests. The second force, client demand, stems from the fact that some investors are beginning to place a higher value on non-economic factors of companies. This concept is not new, as the first socially responsible mutual fund (Pax World Fund) was released in 1971, one of the first socially responsible indexes was launched in 1990 (Domini 400 Social Index), and the World Bank issued their first labeled green bond in 2008. However, as of 2019, over 2,500 companies have signed the Principles for Responsible Investment initiative. The PRI initiative houses the concept of ESG, and the large number of companies signing the initiative indicate that companies are recognizing the importance of ESG in their investor's motivations. Finally, the third force is regulation. Since 1996, it is estimated that over 400 new policies or policy revisions of responsible investment regulations have occurred. Therefore, companies are facing a blend of economic, client, and regulatory pressures to encourage them to be responsible across the entire ESG field as a responsible investment company.

Another indicator that ESG is gaining in popularity is the number of providers that quantify a company's ESG performance. While there are many providers available, the three providers that will be discussed here are: MSCI, Refinitiv, and Sustainalytics. The first provider, MSCI (formerly known as Morgan Stanley Capital International and MSCI Barra), is built around 37 ESG key issues (MSCI, 2019). These 37 issues fall under three pillars (environment, social, and governance), or the 37 issues can be split into 10 themes. Table 10.8 summarizes the 3 pillars, 10 themes, and 37 key issues.

In order to quantify a company's performance in the 37 ESG key issues, MSCI utilizes specialized datasets (from the government, NGOs, and various models), company disclosure records (10–K, sustainability reports, proxy reports), and media sources (global/local news sources, government, NGOs). Through daily monitoring and quality review of the data, MSCI provides a rating of a company's ESG performance. The rating scale is AAA through CCC.

The second provider discussed here that quantifies a company's ESG performance is Refinitiv. Refinitiv is owned by Blackstone Group LP and Thomson Reuters. Like MSCI, Refinitiv has broken down ESG into 10 themes, but has over 400 metrics underneath these 10 themes (Refinitiv, 2019). Refinitiv's 10 themes under the 3 ESG pillars are:

TABLE 10.8 •

MSCI Pillars, Themes, and Key Issues of ESG (MSCI, 2019)

Pillar	Theme	Key Issues
Environmental (E)	Climate change	Carbon emissions, product carbon footprint, financing environmental impact, climate change vulnerability
	Natural resources	Water stress, biodiversity & land use, raw material sourcing
	Pollution and waste	Toxic emissions & waste, packaging material & waste, electronic waste
	Environmental opportunities	Opportunities in clean tech, green building, and renewable energy
Social (S)	Human capital	Labor management, health & safety, human capital development, supply chain labor standards
	Product liability	Product safety & quality, chemical safety, financial product safety, privacy & data security, responsible investment, health & demographic risk
	Stakeholder opposition	Controversial sourcing of work
	Social opportunities	Employee access to communications, access to finance, access to health care, opportunities in nutrition & health
Governance (G)	Corporate governance	Board, ownership, pay, accounting
	Corporate behavior	Business ethics, anti-competitive practices, tax transparency, corruption & instability, financial system instability

- Environmental (three themes): resource use, emissions, innovation
- Social (four themes): workforce, human rights, community, product responsibility
- Governance (three themes): management, shareholders, corporate social responsibility strategy

The two examples of metrics that fall under the human rights theme under the social pillar are: 1) child labor controversies and 2) human rights controversies. In order to quantify these two metrics, the number of controversies published in the media linked to these two issues is counted. A unique feature of Refinitiv is what they call an "ESG controversies score." This additional score, which is considered at the same level as the three ESG pillars, identifies and magnifies controversies surrounding any of the ESG metrics. The controversy score is applied as a negative, or a discount, so it decreases a company's ESG overall score.

Refinitiv tracks media coverage of the 400+ metrics, company annual reports, company websites, NGO websites, stock exchange files, corporate social responsibility (CSR) reports, and news sources to collect data for each company. This data is checked for quality through several filters including interrelated data and variance within a year. Sample audits are performed on a daily basis while Refinitiv management provides monthly "deep dive" reviews to ensure that the data generated is representative of what companies are actually doing. Refinitiv then assigns both a percentile rank score (100–0) and a letter grade (A+ to D-) for a company.

The third provider discussed here that quantifies a company's ESG performance is Sustainalytics. Sustainalytics is a private firm and takes a slightly different approach than MSCI and Refinitiv. Instead of providing a score, rank, or grade, Sustainalytics places a company in one of five ESG risk categories: negligible risk, low risk, medium risk, high risk, and severe risk. These five risk categories, or ESG Risk Ratings, provide a magnitude of whether the company is properly managing any risk associated with ESG, where a low risk indicates that the company is successful with achieving ESG while a high risk indicates that a company is not successful with achieving ESG.

Sustainalytics uses three building blocks to evaluate a company: corporate governance, material ESG issues, and idiosyncratic ESG issues. The foundation of the evaluation rests on corporate governance, with material ESG issues and idiosyncratic ESG issues built on top of the corporate governance. Material ESG issues are applied for all companies, while the idiosyncratic ESG issues are only activated when a company has a very unique issue. Table 10.9 provides the concepts that fall beneath these three building blocks.

TABLE 10.9

Concepts within Sustainalytics for Assigning ESG Risk Categories (Garz and Volk, 2019)

Building block	Concepts
Corporate governance: six pillars	1. Board and management quality and integrity
	2. Board structure
	3. Ownership and shareholder rights
	4. Remuneration
	5. Financial reporting
	6. Stakeholder governance
Material ESG issues: big six	1. Human capital
	2. Business ethics
	3. Product governance
	4. Carbon – own operations
	5. Occupational health and safety
	6. Data privacy and security
Idiosyncratic ESG issues: a sampling of events	• Marine ports: land use and diversity
	• Metals and mining: employee human rights
	• Agricultural chemicals: society human rights
	• Home furnishings: accounting and taxation
	• Electric utilities: corporate governance

In order to assign a company to a risk category, Sustainalytics follows three steps. First, a company's exposure to the ESG issue is assessed. Next, the management of the exposure and the degree of management is evaluated, and finally, the unmanaged risk is calculated. The data is evaluated daily through assessments of public disclosure reports, media, and NGO reports. Annual exposure assessments are performed and data is refreshed when there are any significant events across the world. The companies in the high- and severe-risk categories have additional safeguards in the evaluation process by Sustainalytics to ensure a high level of quality during the review. Finally, draft reports are sent to every company that is evaluated to get feedback from the company. These layers of evaluation allow for a high level of confidence in the quality of the data being collected and ensure that the final category is appropriate.

While the concept of ESG in corporations has been active for decades, it became mainstream in the late 2010s. Therefore, it is still a dynamic field that could have significant changes in the short term. Yet, even with these changes, the fundamental concept of ESG appears to have moved into standard corporate strategy. Therefore, it is worthy of consideration for Civil Engineering firms to consider developing and adopting an ESG strategy as the nature of Civil Engineering work touches many different lives and types of businesses. This could not only add value to a Civil Engineering firm, but would also increase the social responsibility of the firm.

Example Problem 10.4

The California Public Employees' Retirement System (CalPERS, System) produced a document on governance and sustainability principles; google the phrase "CalPERS' Governance & Sustainability Principles." When examining this document, the Principles for Responsible Investment, or PRI, are listed that were generated from the UN and UNEP. What are the six principles?

The six principles are: (1) We will incorporate ESG issues into investment analysis and decision-making processes, (2) We will be active owners and incorporate ESG issues into our ownership policies and practices, (3) We will seek appropriate disclosure on ESG issues by the entities in which we invest, (4) We will promote acceptance and implementation of the Principles within the investment industry, (5) We will work together to enhance our effectiveness in implementing the Principles, and (6) We will each report on our activities and progress towards implementing the Principles.

10.5   SUMMARY

Chapter 10 has covered a lot of ground. Several certification programs were identified and discussed, including LEED, Envision, and GreenRoads. While these certificate programs generally focus on specific products, companies also have opportunities to explore the economic, environmental, and social aspects of their own practices. This company-driven mind-set was discussed through ESG: Environmental, Social, and Governance. However, these programs are not the only programs available when exploring certification. For example, the Environmental Protection Agency's

(EPA's) Energy Star program has a large impact on residential construction, various commercial and industrial building applications, and energy-efficient products. Also, certification programs such as BREEAM are extremely popular in England and around the world, while Green Globes has a large presence in Canada. These programs, among others, have extensive resources online to help better understand their unique characteristics.

Another concept that warrants a brief discussion is the concept of cost of certification on a project, and whether that cost is recovered due to better performance. Costs of certification include the certification fees themselves, higher potential design costs, waste management on the construction site, costs associated with sustainable materials, and post construction monitoring. However, over the life of a building, saving can be earned from reduced energy usage, lower environmental impact, increased social benefits, among other things. Therefore, it can be easy to "justify" the cost of a green project and it can be easy to prove that green projects are not worth the expense. Some of the example problems and homework problems are designed to provide resources to more fully explore the costs of green projects. In summary, since so many certification programs exist, and the list is only growing, enough owners have found the certification process worth any additional resources.

HOMEWORK PROBLEMS

1. Within the document "The Costs and Financial Benefits of Green Buildings," list the nine potentially promising ways green buildings could help address a range of challenges facing California.
2. Within the document "The Costs and Financial Benefits of Green Buildings," Figure III-2 outlines the average green cost premiums for 31 LEED buildings, what is the average green premium (in percent) of certified, silver, gold, and platinum buildings?
3. Do you think that the premiums associated with LEED construction are worth the extra cost? In order to answer this question, use the format provided under Sidebar 1.2 "Writing a High-Quality Essay."
4. Choose a LEED-certified building in your area and find the score on the USGBC website. List two credits from Table 10.1 that you believe the building achieved, and two that you believe it did not achieve. In order to answer this question, use the format provided under Sidebar 1.2 "Writing a High-Quality Essay."
5. Within the document "Use of the Envision Sustainable Infrastructure Rating System for Water Infrastructure," what were the six goals for the 26th Ward Wastewater Treatment Plant in regard to Envision?
6. Within the document "Use of the Envision Sustainable Infrastructure Rating System for Water Infrastructure," what are the two requirements of the County of Los Angeles Department of Public Works for contractors?
7. What is one similarity and one difference between LEED and Envision? Pick one similarity and one difference, and use the format provided under Sidebar 1.2 "Writing a High-Quality Essay" to build your argument.

8. Within the document "Greenroads: A sustainability performance metric for roadway design and construction," how does Greenroads define sustainability? This one sentence also defines the three laws within the definition, what are the three laws? Your answer only needs to be four sentences long.

9. Within the document "Greenroads: A sustainability performance metric for roadway design and construction," Greenroads is defined as a project-based system. Which two steps in the processes are specifically addressed within the system boundaries, and what six items associated with roadways are considered in specific ways?

10. Do you think that Greenroads is significantly better than LEED or Envision certification? Use the format provided under Sidebar 1.2 "Writing a High-Quality Essay" to build your argument.

11. Within the document "CalPERS' Governance & Sustainability Principles," what are the five core issues that they believe have a long-term impact on risk and return?

12. Within the document "CalPERS' Governance & Sustainability Principles," the board quality section has a list of board responsibilities, which include a discussion on corporate culture. What five cooperate culture areas should the board have an active role in setting?

13. Within the document "CalPERS' Governance & Sustainability Principles," the regulatory effectiveness section has a discussion on sustainable policy framework, which includes productive labor practices and environmental risk factors. Do you think that the discussion in this section is adequate, or are core concepts of sustainability missing? Use the format provided under Sidebar 1.2 "Writing a High-Quality Essay" to build your argument.

REFERENCES

Anderson, J., Muench, S. Sustainability Trends Measured by the Greenroads Rating System, Transportation Research Board, No. 2357, pp. 24–32, 2013.

Bowles, E., Venner, I., Stanford, D., Sheppard, C. Use of the Envision Sustainable Infrastructure Rating System for Water Infrastructure, Fact Sheet, WSEC-2017-FS–018, Water Environment Federation, WEF ENVISION Task Force, Institute for Sustainable Infrastructure (ISI), 2017.

CalPERS. CalPERS' Governance & Sustainability Principles, California Public Employees' Retirement System, September 2019.

Envision. *Envision: Sustainable Infrastructure Framework Guidance Manual*, Institute for Sustainable Infrastructure, Washington, DC, 3rd Edition, 2018.

Garz, H., Volk, C. The ESG Risk Ratings: Moving up the Innovation Curve, White Paper – Volume 1, Sustainalytics, updated January 2019.

Greenroads. Manual v1.5, Greenroads International, 2011.

Greenroads. Rating System V2 (Free Sample Download), Greenroads International, April 2019.

Kats, G., Alevantis, L., Berman, A., Mills, E., Perlman, J. The Costs and Financial Benefits of Green Buildings, A Report to California's Sustainable Building Task Force, Capital E, October 2003.

Kocmanová, A., Šimberová, I. Determination of Environmental, Social, and Corporate Governance Indicators: Framework in the Measurement of Sustainable Performance. *Journal of Business Economics and Management*, 2014, 15(5), 1017–1033.

LEED. LEED v4.1, Building Design and Construction: Getting Started Guide for Beta Participants, U.S. Green Building Council, July 2019.

Lew, J., Anderson, J., Muench, S. Informing Roadway Sustainability Practices by Using Greenroads Certified Project Data, Transportation Research Board, No. 2589, pp. 1–13, 2016.

Muench, S., Anderson, J. Greenroads: A Sustainability Performance Metric for Roadway Design and Construction, TNW2009-13, Washington State Department of Transportation, 2009.

MSCI. MSCI ESG Ratings Methodology, Executive Summary, MSCI ESG Research, September 2019.

Pittenger, D. Evaluating Sustainability of Selected Airport Pavement Treatments with Life-Cycle Cost, Raw Material Consumption, and Greenroads Standards, Transportation Research Board, No. 2206, 2011, pp. 61–68.

PRI. What is Responsible Investment, An Introduction to Responsible Investment, Principles for Responsible Investment (PRI), UNEP Finance Initiative and UN Global Compact, 2019.

Refinitiv. Environmental, Social and Governance (ESG) Scores from Refinitiv, RE965397/7–19, June, 2019.

Rodrigues, M. Applying the ISI Envision Checklist, Post project analysis of the Alder Street Active Transportation Corridor Project. *City of Eugene, Public Works Engineering*, July 19, 2013.

UN. Transforming our World: The 2030 Agenda for Sustainable Development. Resolution adopted by the General Assembly on 25 September 2015, Seventieth session.

11 Tomorrow's Sustainability

The principal goal of education in the schools should be creating men and women who are capable of doing new things, not simply repeating what other generations have done.

Jean Piaget

While the concept of sustainability has been in existence for millennia, it is only in the past 30 years that there has been significant movement toward the qualification and quantification of sustainability. From a qualitative standpoint, the UN, the ASCE, Oxfam, and various other groups have taken strides to defining sustainability. The concept of three pillars of sustainability is common through this work, with the three pillars being economic, environmental, and social. From a quantitative standpoint, concepts such as Life Cycle Cost Analysis, benefit/cost ratio, Life Cycle Assessment, and ecological footprint have captured economic and environmental metrics of sustainability, and the field of social metrics is young but developing rapidly. Through all of this work, however, many would agree sustainability is not a straightforward issue, and overall, is a relatively undeveloped field. The advancement of sustainability in civil engineering is not an exception. In addition, many believe that in order to truly succeed in implementing sustainable practices in civil engineering, a paradigm shift will need to occur with innovative and applicable solutions.

When reviewing the plethora of resources available that discuss the future of sustainability, three stood out as both thoughtful and comprehensive in their analysis. These three documents were the UN's "2030 Agenda for Sustainable Development" (UN, 2015), Global Reporting Initiative's "Sustainability and Reporting Trends in 2025" (GRI, 2015), and the World Conservation Union's "Future of Sustainability" (IUCN, 2006). These three documents are broken down in the pages that follow to discuss how sustainability is not a straightforward issue, how it is underdeveloped, and how a paradigm shift will need to occur moving forward.

11.1 SUSTAINABILITY IS NOT A STRAIGHTFORWARD ISSUE

Starting with the UN, their 2030 Agenda contained 17 goals, 4 of which are directly related to civil engineering. The four goals directly related to civil engineering are:

- Goal 6 – ensure availability and sustainable management of water and sanitation for all
- Goal 9 – build resilient infrastructure, promote inclusive and sustainable industrialization, and foster innovation
- Goal 11 – make cities and human settlements inclusive, safe, resilient, and sustainable

- Goal 17 – strengthen the means of implementation and revitalize the Global Partnership for Sustainable Development

Just looking at these four goals brings a host of questions. Can the water and sanitation system currently installed in high-income (i.e. developed) countries simply be replicated and placed into low-income (i.e. undeveloped) countries, as outlined in the sixth goal? In the ninth goal, how can resilient infrastructure be built while achieving all three pillars: economic, environmental, and social? How can we as civil engineers incorporate safety into all of our designs, whether buildings or roads, as charged by goal 11? These are not easy, nor trivial, questions. But as civil engineers, following ASCE's code of ethics, we must "use [engineer's] knowledge and skill for the enhancement of human welfare and the environment." All of these discussion points can also be indirectly addressed when considering construction management principles, and how we execute civil engineering projects.

More challenges are presented by GRI, which presents 10 sustainable economic challenges, 3 of which are directly related to civil engineering:

- Challenge 1 – shortage of raw materials
- Challenge 3 – reduce waste and ecosystem contamination
- Challenge 7 – define regional sustainable development plans

Here, things appear a bit more straightforward. In regard to the first challenge, civil engineers are well aware that so many of our tools are finite resources; this is why fly ash is substituted for Portland cement in Portland Cement Concrete, steel is one of the most recycled materials on earth, and Reclaimed Asphalt Pavement is substituted for aggregate and asphalt binder in asphalt mixtures. For the second challenge, the United States has made tremendous strides in reducing ecosystem contamination with the passage of the Clean Air Act in 1963 and the Federal Water Pollution Control Act Amendments of 1972, which led to the Clean Water Act of 1977. Yet, these examples provide a contradiction of sorts. If civil engineers were truly achieving a reduction in raw materials by using fly ash, recycled steel, and RAP, would GRI continue to list the shortage of raw materials as the number one challenge? Another question comes from the seventh challenge. Regional planning is rarely performed by civil engineers, but instead it is usually directed by government-appointed planning commissions. Therefore, how can civil engineers use their skills to introduce options to these commissions, armed with the knowledge of the economic, environmental, and social perspectives of the long-term projects that will be constructed from the regional plan? Again, these discussion points can also be viewed through the lens of construction management principles, and how we plan and administer civil engineering projects.

The situation does not become any simpler when examining the 13 regulating and cultural services put forth by IUCN, 4 of which are directly related to civil engineering:

- Service 1 – air quality regulation
- Service 4 – water regulation

- Service 5 – erosion regulation
- Service 6 – water purification and waste treatment

Here, we see similar themes to the UN and GRI, with air quality and water taking three of the four services. The fifth service, however, is new, but no less important. Erosion takes many forms, from runoff of construction sites to high-quality topsoil being washed away on cleared farmland and the degradation of coastlines. Therefore, between the UN, the GRI, and IUCN, all five of application areas of civil engineering have been covered, with discussion and challenges in environmental, geotechnical, structural, transportation, and construction management. In addition, the development of sustainable certification programs and business frameworks allow for the implementation of sustainable concepts that span all of the UN, GRI, and IUCN objectives.

But the challenges reviewed were only those directly related to civil engineering and did not weigh in on the "soft" issues, such as poverty, hunger, climate change, wealth inequality, gender inequality, social conflict, human rights, education of workers, or anti-corruption policies. Therefore, in addition to the challenges associated directly with civil engineering, challenges that fall outside of the direct umbrella of civil engineering must be faced. The bright side to these challenges, both within and outside of civil engineering, is that tremendous opportunities are available for improvement of existing practices and the development of new practices to move toward a more sustainable future.

11.2   SUSTAINABILITY IS AN UNDEVELOPED FIELD

While there has been progress in both the economic and environmental pillar of sustainability and there is an acknowledged need for progress in the social pillar, there is another existing opportunity that allows for impactful opportunities for advancement of sustainable practices. As asked in Chapter 11.1, is it appropriate for sanitation systems in high-income countries to simply be placed in low-income countries? These highly complicated and expensive sanitation systems require a complex collection system to move the wastewater from buildings to the treatment plant, and then the wastewater must be treated through a multistep process that includes pretreatment, primary treatment, secondary treatment, tertiary treatment, and disinfection. Finally, the fully treated water needs to be returned to the ecosystem. All of these stages require extensive equipment and chemicals in order to execute. Interestingly, some of our youngest, and unpaid, workers are providing opportunities to solve these problems in low-income countries using more appropriate solutions.

Engineers Without Borders USA is a student group that has chapters at many universities across the United States. Generally, small groups of students with a faculty member or two travel to a low-income country in order to complete an engineering-based volunteer project. For example, two groups of civil engineering students went to Costa Rica to install bathrooms in a new computer center at a small school in San Juan de San Isidro (Texas A&M, 2010). Obviously, the students would not be building a full-scale wastewater distribution system and wastewater treatment plant, so they instead installed a small septic tank with an on-site leach field. The total cost for the plumbing material was just under $3,700, a cost achievable for a low-income community.

This is just one example of the over 650 community-driven projects in over 40 countries around the world headed up by Engineers Without Borders USA. However, just like wastewater treatment technology from high-income countries should not be applied to low-income countries, the wastewater project in Costa Rica may not work in other low-income countries. The challenges associated with students building tailor-made projects for specific communities, while still utilizing engineering principles learned in the classroom, show the possibilities of future potential developments in the field of sustainability.

11.3 PARADIGM SHIFT REQUIRED FOR SUSTAINABILITY

To provide an industry perspective, the University of Arkansas invites representatives from different fields of civil engineering to speak to the sustainability class. About four years ago, during the question and answer portion at the end of the class, the invited company representative was asked if their company would ever consider environmental impact on a level playing field as an economic impact. The answer was a flat out "no." While not surprising, many people answer that question more along the lines of "both the environment and economic impact are important, but of course, if we are to stay in business, the economic case must be strong." Credit should be given to the company representative for answering truthfully. Four years later, a different invited speaker shared that while the economic impact to customers is the primary driving force, sustainable solutions have become more commonplace due to local government requirements regarding environmental impacts of development projects. This is interesting as it indicates a positive trend toward sustainability, but it also highlights the work that needs to be done in changing how we, as civil engineers, approach our work.

GRI agrees that progress needs to be made in the area of environmental and social sustainability. In its report, GRI discusses how companies must be held accountable, and that the business decision makers need to take sustainability issues into account more profoundly. The IUCN agrees, calling for "new concepts, new thinking." The challenge for incorporating the environmental pillar into our civil engineering design process will be difficult, but chances are, incorporating the social pillar will be even more difficult. Therefore, existing employees of our civil engineering companies and agencies must continue to move toward economic, environmental, and social improvements in our design and execution of infrastructure projects.

Our future rests with the students of today and tomorrow. The primary purpose of this book is to provide tools for students to begin tackling these complicated, non-straightforward issues. However, the book also strives to embrace the wide open field of sustainability, change the mind-set of graduating civil engineers, to demonstrate that the economic pillar of sustainability is only one-third of the picture, and to convince students becoming professionals the environmental and social pillars of sustainability are just as important as the economic pillar in civil engineering. All of these changes will probably not happen today, but they need to happen. Hopefully, the reader now has a more complete toolbox to tackle and address sustainability, so that tomorrow, the changes will happen, and the world will be a better place for all of the inhabitants.

REFERENCES

GRI. Sustainability and Reporting Trends in 2025, Preparing for the Future. *Global Reporting Initiative*, May 2015.

IUCN. The Future of Sustainability, Re-thinking Environment and Development in the Twenty-first Century. *The World Conservation Union, Report of the IUCN Renowned Thinkers Meeting*, January 29–31, 2006.

Texas A&M. Implementation Phase II: Wastewater System, Building Construction, and Wiring. *Engineers Without Boarders Texas A&M University Student Chapter, Post Implementation Report, Document 526*, 2010.

UN. Transforming our world: the 2030 Agenda for Sustainable Development. Resolution adopted by the General Assembly on 25 September 2015, Seventieth session.

Index

.